Who Really Invented the
COTTON GIN?

Who Really Invented the COTTON GIN?

UNRAVELING THE MYSTERY AND FOLKLORE OF A CRITICAL AGRICULTURAL INNOVATION

Wesley F. Buchele, PE, PhD

William D. Mayfield, MSAE

Buchele Associates, LTD, First Edition

February 2016

ISBN: 978-1-5303-1178-1

CONTENTS

ABOUT THE AUTHORS

Wesley F. Buchele is a professor emeritus at Iowa State University and consulting engineer. Before retiring in 1989, Buchele taught farm machinery design for forty-three years. He holds twenty-three patents. The most important are: (1) the rotary threshing and separating cone-cylinder used in all modern rotary grain combines, and (2) the large round baler that is seen everywhere. Early in his professional life, Buchele spent three years (1948–1951) researching cotton mechanization in the Mississippi Delta.

William D. Mayfield spent his entire professional career working on cotton mechanization. He retired as the national program leader for cotton ginning technology for the USDA.

The Mayfield Cotton Engineering Award is sponsored by the John Deere Foundation to honor him, an extension agricultural engineer. He devoted his entire life to applying engineering fundamentals to the production, harvesting, and ginning of cotton.

Buchele and Mayfield have used their technical and investigative expertise to determine just who really did invent the first practical, cylindrical, continuous-flow, rip-saw-toothed cotton gin.

ACKNOWLEDGMENTS

I am indebted to Ruth Meyer, my former secretary at Iowa State University; to professor emeritus Stephen J. Marley, my colleague for twenty-six years in the Department of Agricultural and Biological Engineering at Iowa State, for his thorough review of the manuscript; to my daughter Marybeth Buchele, for reading and correcting various drafts; to Benjamin Hucker for his technical assistance during the writing process; to Nicholas Berry for his assistance with the cotton production graphs; to Angela Lakwete, associate professor of history at Auburn University, for her encouragement and for answering my many questions; to Warren E. Garner, retired director of the U.S. Cotton Ginning Research Laboratory in Stoneville, Mississippi, for supplying copies of historical documents; to professor emeritus Absalom Snell of the Department of Agricultural Engineering, Clemson University, South Carolina, for squiring me around the Clemson campus and for his encouragement; to professor emeritus Clarence Johnson of Auburn University for introducing me to a modern cotton gin factory; and to Linda Moore, associate director of the Burton Cotton Gin and Museum in Burton, Texas, for enthusiastically demonstrating the history of cotton ginning. I also want to thank Robert Schafer and Clarence Johnson, my former graduate students at Iowa State, for volunteering to help with the publishing of this book. They commanded me to write and write and continue to write! Yes, sirs.

— *Wesley F. Buchele*

ACKNOWLEDGMENTS

As a fresh engineering graduate and a new employee of the Cotton Ginning Research Laboratory at Stoneville, Mississippi, in 1966, I became aware of the history of cotton ginning by reading the books written only a few years earlier by Charles A. Bennett. Mr. Bennett was the laboratory director until his retirement in 1962.

As a member of the ASAE Historic Commemoration Committee, I led the investigation of the invention of the cotton gin by following up on Mr. Bennett's references. This committee agreed that Henry Ogden Holmes probably invented the first gin, but conclusive information was not available to discredit Eli Whitney. Therefore, we recognized both Whitney and Holmes on the commemorative plaque erected in Savannah, Georgia.

I primarily acknowledge my coauthor, Dr. Wesley Buchele, for his enthusiastic research into court records and other documents that contributed significantly to knowledge concerning the invention of the cotton gin and the invention's impact on the world economy.

— *William D. Mayfield*

*W*e also are both indebted to the many people who gave and collected oral histories of the cotton gin and to the authors of the many books about the folklore and history of cotton and cotton gins that we consulted. We have paraphrased their contributions to history and referenced them in our book.

Finally, we thank Lee Klancher, Publisher, Octane Press, for providing expertise and arranging for experts to bring the book to publication; Tom Heffron, graphic designer, for design of the book cover and for the manuscript layout design; and Leah Cochenet Noel, copy editor, for an outstanding editing job. She expertly turned the original text into that which is easily understood. Thanks also go to Danielle Magnuson, for giving the final text a very thorough proofread and catching lingering typos and more.

— *Wesley F. Buchele*
William D. Mayfield

EDITOR'S NOTE

In compiling images to go with this book, Greg Smith, director of archival investigations at History Picquette LLC, did much research and work to find a historical sketch or photo of Hodgen Holmes and sadly wasn't successful. But he did find some images from Dr. Buchele's 2006 trip to South Carolina, where Buchele did further research on this book and hunted for old cotton gins in Fairfield County barns. Those photographs appear in chapter 2.

PREFACE

The Revolutionary War was over. In the aftermath, Americans quickly shed the commercial restrictions imposed by the British government on the colonies. They had suffered an embargo of industrial goods for eight years. They now began to solve their industrial and agricultural problems for themselves.

Southern planters had been growing green-seeded, short-staple cotton for years in their gardens. They finger-ginned, spun, and wove it into homespun fabric for their and their slaves' use. Woolen clothing was itchy, and only the rich could afford linen, so southerners wore cool cotton during the hot summer days. However, to develop cotton into an industry, the first order of business was to invent a cotton gin, and invent they did—two types of gins, in fact. Now most Americans who know anything about a cotton gin would tell you Eli Whitney first came up with the invention. But is that true?

As early as 1784, southern blacksmiths began manufacturing continuous-flow, rip-saw-toothed cotton gins. Ten years later, in 1794, Eli Whitney invented and began manufacturing a batch-type, bent-needle-toothed cotton gin.

This book details and reexamines the official history of the cotton gin. It combines southern folklore with historical documentation and an engineering analysis of saw gins, based on both patent records and surviving examples. Also presented are American cotton shipping data to demonstrate the growth of the cotton industry, as well as the patents, lawsuits, and legal injunctions that complicate the history of this important technology.

As authors, our goal was to determine who really invented the first practical, modern rip-saw-toothed cotton gin. And specifically, who invented the cylindrical, continuous-flow, rip-saw-toothed cotton gin that has ginned nearly all of the green-seeded cotton grown in the world since about 1784. Read on to find out what we learned.

— *Wesley F. Buchele*
William D. Mayfield

CHAPTER 1

Introduction

Many schoolchildren across the country learn the conventional history about the cotton gin—that young Eli Whitney was the first to invent it (in 1793) and patent it (on March 14, 1794). However, southern folklore tells a different story—that a young blacksmith named Henry Ogden Holmes, later known as Hodgen Holmes, patented the first practical cotton gin. Holmes, a resident of Hamburg (some sources say Bamberg), South Carolina, applied to the U.S. War Office for a "caveat of invention" for his gin in 1787. Before the establishment of the U.S. Patent Office, the caveat of invention was the only means inventors had to protect their ideas.

Holmes' caveat of invention was granted on March 14, 1789, with a life of five years. George Washington himself would have signed it. However, because this document has never been found in modern times, Holmes' caveat of invention may have never existed.[1] Additional folklore states that, in 1787, a prosperous planter named James Kincaid employed Hodgen Holmes to bring his newly invented rip-saw-toothed cotton gin to his plantation and operate a commercial cotton gin in Kincaid's gristmill. Some say that Kincaid also helped Holmes submit his application for a caveat of invention to the U.S. War Office.

Nine years later, Hodgen Holmes submitted an application to the U.S. Patent Office to patent a continuous-flow rip-saw-toothed cotton gin. This patent was granted on May 12, 1796. However, on November 6, 1793, the U.S. Patent Office received a patent application from Eli Whitney for a batch-type, bent-needle cotton gin. Whitney was granted his patent on March 14, 1794. In fact, because the application had been received on November 6 of the previous year, the patent was back-dated to 1793. The life of the patent was fourteen years, so it expired on November 6, 1807. Aware of this, Whitney petitioned Congress, unsuccessfully, in 1807 and 1811 to extend his patent.

Volume III of the *American Historical Review* published in July 1898 states on page 93 that some people believe that Hodgen Holmes was the true inventor of the saw cotton gin and that Eli Whitney stole the invention from him.

Whitney's patent covered an impractical, batch-type cotton gin equipped with bent-needle teeth and a rectangular, slotted wooden breastwork that lay on top of the bent, needle-toothed cylinder. While the patent drawings of Whitney's gin do not show a triangular breastworks with an apron in the front, all models of Whitney's gins that exist to this day have aprons that would receive the seeds as they were stripped from the fiber by the breastworks. Slots sawn in the breastwork and apron matched the rows of bent needles on the rotating wooden cylinder.

The operation of Whitney's gin was a batch process. A limited quantity, or batch, of unginned cotton lint, containing seed attached in the lint, was placed in a bin, manually fed through Whitney's gin, and collected in an output bin. The output bin then contained cotton lint without seed and some locks of cotton still containing seed that needed to be separated and recycled through the gin until the seeds were separated from the cotton lint. Consequently, using Whitney's gin was labor intensive, time consuming, and inefficient. The operation of the bent nails on the gin's rotor did not efficiently separate all the cotton lint from the cotton seed in one pass through.

Hodgen Holmes' patent described a continuous-flow cotton gin that was equipped with circular metal blades that were spaced between vertical metal ribs. Holmes' gin did not need to be emptied of accumulated seeds, and the circular metal blades were both easier to manufacturer and more reliable in operation. Nevertheless, in a trial that ended on November 6, 1802, Judge William Stephens nullified Holmes' May 12, 1796, patent.[2] In a further judgment, Judge William Johnson stated that Holmes' gin was "merely an adaptation" of Whitney's bent-needle gin. Finally, on December 19, 1806, Johnson decreed a permanent injunction against Holmes' estate for the manufacture of rip-saw-toothed cotton gins. In effect, Judge Johnson awarded Holmes' patent to Whitney, and Holmes was forced to buy a license from Whitney in order to manufacture his own invention.

All told, Whitney and his associates brought more than twenty-eight lawsuits against the manufacturers and vendors of cotton gins in Savannah, Georgia, alone and an unknown number of lawsuits in other cities of the South. The first twenty-three trials ended in non-suits, dismissed suits, or not-served suits. Whitney received plaintiff verdicts in only four of his last five patent suits in Savannah.

Some, including the American Society of Agricultural and Biological Engineers, have called Holmes an improver of Whitney's cotton gin. But we call him the real inventor of the cylindrical, continuous-flow, rip-saw-toothed cotton gin. And we are not alone.

Unfortunately, most of the authors who have written about the history of the cotton gin did not have the technical expertise to appreciate the important differences between the two early cotton gin designs. None of the judges or the authors, except the engineers, gave any attention to the automatic disposal of the ginned seed in Holmes' invention.

Holmes' gin teeth were cut in the rim of steel saw blades and lasted a lifetime while Whitney's gin had bent needle teeth that twisted in or pulled out of the wooden cylinder. Holmes' gin continuously operated as the seed roll automatically carried the ginned seed around and dropped seeds out of holes in the bottom of the hopper.

Whitney's gin was a batch-type gin. When the hopper became engorged with ginned seeds, the operator stopped it and removed all the ginned cotton and the unginned cotton in the hopper and spread them out on a table. The unginned cotton was sorted out of the mass cotton by hand and returned to the hopper with new unginned cotton.

What is most troubling about Holmes' fight for his patent was that Whitney and the judges never understood the continuous operation of his gin. Yet, the judge gave Whitney the Holmes' gin patent in 1806, which another judge had nullified in 1802.

At the turn of the twentieth century, Daniel Tompkins, professor of mechanical engineering at Clemson University, promoted the production of cottonseed oil, which contributed to the revitalization of the southern economy. He also reconstructed three full-size cotton gins that were in use during the 1790s. These reproduction cotton gins are now on display at three colleges in South Carolina. Another engineer, Charles A. Bennett, the principal agricultural engineer of the Cotton Ginning Section of the Agricultural Research Service, USDA, from the 1930s to the 1960s, promoted the development of higher-capacity gins and wrote the definitive book on cotton gin engineering.[3]

Because fire destroyed the U.S. Patent Office on December 15, 1836, Tompkins and Bennett had to rely on other documents from 1793, 1794, 1796, 1802, 1804, and 1808 to decide if Holmes had the first patent for the cotton gin. These documents are still available and are stored in the U.S. District Court House in Savannah, Georgia, and in the Georgia State Archives in Atlanta. In addition, Kaller Historical Documents, Inc., of Marlboro, New Jersey, currently owns a copy of the title page of Holmes' 1796 patent, and Whitney's patents (including specifications, descriptions of the gin and their operation, drawings, and claims) are reproduced in the appendix of Tompkins' book.

A Brief History of Cotton Ginning

Long before the patent fight between Whitney and Holmes erupted in the post-Revolution colonial era, using cotton for fabric became part of civilization. Ancient mummies wrapped in cotton fabric have been found in the deserts of Pakistan, India, China, Egypt, and Peru.[4] Columbus, Cortez, Pizarro, and Magellan reported seeing various uses of cotton as they explored the western hemisphere.[5]

Commercial cotton has long come from two different kinds of the crop: green-seeded and black-seeded cotton. Most green-seeded cotton varieties are grown and harvested as annual shrubs in colder climates and are killed by frosts in the fall. Most black-seeded cotton is grown as small perennial trees in the warm, frost-free areas of the world.

Green-seeded cotton with a staple length of ¾ to 1¼ inches is often called upland cotton or short-staple cotton. These varieties grow well on the uplands of India, Pakistan, China, and the

Americas.[6] While green-seeded varieties are hardier and more prolific, one disadvantage is that the short fibers hold tenaciously onto the seeds and resist separation. Before the invention of the cotton gin, the lint had to be laboriously finger-ginned—that is, the seeds were separated from the fibers by hand, at a rate of about one pound of lint per person per day. This made green-seeded cotton very expensive. The green seeds, after ginning, still carry a fuzzy coat. The finger-ginned quota of slave families of the mid-1700s was four pounds of lint per week.[7]

Black-seeded cotton, with its silky, long staples (longer than 1¼ inches), is often called long-staple cotton or Sea Island cotton. Rather than ginning, early Mesoamericans used selective breeding so the seeds are only loosely attached to the fibers.[8] As a result, black-seeded cotton can be ginned by pinching the fibers and squeezing the bald seeds out. Black-seeded cotton is grown in frost-free, warm regions, such as Egypt and Arizona. Because it is more difficult and expensive to grow than green-seeded cotton, black-seeded cotton is found only in luxury clothing.

Nearly every culture that produced green-seeded cotton also invented a rod-type cotton gin.[9] The simplest rod-type gin consisted of a smooth wooden rod, or eventually a wrought-iron rod, that was rolled on a flat surface of wood or stone using the feet or hands. When the rod was rolled on the trailing fibers against the seeds in the cotton, the rod pinched the seeds forward, separating them from the fibers. It required great dexterity to operate the rod against the tenacious, fuzzy seeds without crushing them and spilling their oil onto the lint.[10] Rod-type gins were in use in Asia, Africa, and the Americas before recorded history.

The next improvement in cotton ginning technology was the churkha, a hand-operated double-roller gin that was invented in

A rod-type cotton gin. *Lorraine Kuehnel*

India in about 300 BC.[11] It was made of two teakwood or wrought-iron rollers that looked much like the wringer of an old-fashioned washing machine. The rollers of the churkha were typically less than a foot long; one had the diameter of a nickel and the other of a dime.[12] As the rollers were turned, they gripped the fibers and pulled them forward while simultaneously pinching the seeds rearward and separating them. Skilled ginners could pull green-seeded cotton fibers cleanly through the pinch point of the rollers. However, an unskilled ginner could crush the seeds, pressing oil and broken seed hulls into the fibers. The oil-soaked fiber was deeply discounted in the marketplace. The broken seed hulls had to be hand-picked, by those who became known as cotton pickers, from the lint, or the lint had to be beaten with sticks to eliminate the debris.

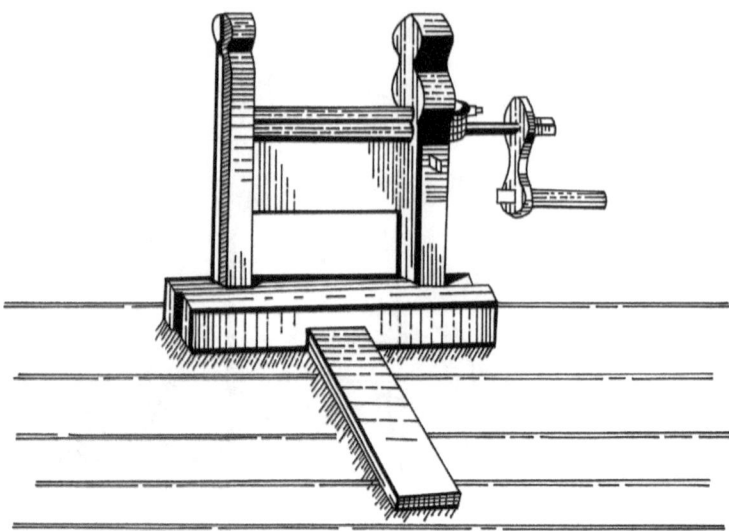

A two-roller hand-cranked churkha gin. It was eventually developed into a foot-treadle machine in the Caribbean. *Lorraine Kuehnel*

Churkha gin

In this sketch, Turks are shown operating a churkha cotton gin.
Authors' collection

By the year 1300, inventors in India and China were fabricating numerous types of hand- and foot-operated rod-type cotton gins for short-staple, green-seeded cotton.[13] Many cultures also built lint-cleaning devices that vigorously fluffed the lint to shake out debris. In particular, India produced master spinners and weavers. They spun and wove finger-, rod-, and churkha-ginned green-seeded cotton lint into fine, colorful fabrics and exported them to the West. Many current names for textiles come from the Indian and Middle Eastern cities that specialized in them or come from local descriptive words. Some examples are calico (from Calcutta), muslin (from Mogul), dungaree (from Danger Kilda near Mamba), and khaki (meaning "dust-colored").[14]

Originally, hand cranks were located on the opposite ends of the churkha's two rollers. One operator cranked a roller with one hand and fed cotton into the pinch point with the other hand, while a second operator turned the other roller and gathered the cleaned lint coming out the back side of the churkha. Later, a pair of wooden gears was attached to one end of the rollers. A single operator could then crank the rollers with one hand and feed the gin with the other.

By the mid-1700s, the English demand for cotton encouraged Caribbean carpenters to build roller gins that were suitable for black-seeded cotton. The traditional churkha ginned thirty to fifty pounds of black-seeded cotton per day. They attached a treadle to each roller, so the operator could run the gin by his or her feet, allowing the operator to feed the gin with both hands. Because it was more efficient, the foot-powered churkha gin became the gin of choice for long-staple cotton. Caribbean carpenters invented and manufactured multi-roller barrel gins and eventually eliminated both hand cranks and gears.

Churkha gins were imported into South Carolina from the Bahamas in 1776 and were used at first unsuccessfully for ginning green-seeded cotton lint.[15] However, in 1785, black-seeded cotton was brought from the Bahamas to St. Simons Island off the coast of Georgia.[16] In 1786, Frank Levitt planted black-seeded cotton on Sapelo Island off the coast of Georgia.[17] From there, black-seeded cotton was profitably cultivated in tidewater lands from Delaware to Florida, and, with the use of barrel gins, it quickly became a cash crop.

In 1778, loyalist and black-powder manufacturer Dr. Joseph Eve fled Philadelphia and began growing black-seeded cotton in

Churkha gin

Two women operating a churkha cotton gin in the late 1800s.

Photography Collection, Miriam and Ira D. Wallach Division of Art, Prints and Photographs, The New York Public Library, Astor, Lenox, and Tilden Foundations

the Bahamas. In 1786, Eve invented a self-feeding roller cotton gin (with two sets of double iron rollers, ⅝ inch in diameter and 30 inches long). These barrel gins could be powered by water, wind, or animals and had a capacity of 300 pounds of black-seeded cotton or 150 pounds of green-seeded cotton per day. After the war, in 1786, Eve emigrated to Augusta, Georgia, and began manufacturing barrel gins. Eve's gins successfully competed with both locally made and commercially manufactured saw gins for ginning green-seeded cotton until the 1820s.[18]

Cotton Production in America

The native people on the island of San Salvador wore fabric made from Sea Island cotton when they paddled out to meet Columbus on October 12, 1492. They gave the admiral two skeins of cotton yarn in honor of his visit.[19] Columbus carried the cotton skeins and six prisoners back to Spain to prove that he had reached India.[20] He was, of course, mistaken. Some years later, Alar Núñez Cabeza de Vaca found upland short-staple, green-seeded cotton growing wild in Louisiana and Texas.[21]

The Virginia Company, chartered by King James I and financed by British mercantile companies, landed its colonists in Jamestown, Virginia, on May 13, 1607, with the intent of planting green cotton seeds from India, among other ventures. That first year, they planted too late to produce a crop, but next year's crop was more successful.[22] By 1616, the colonists got the hang of growing short-staple, green-seeded cotton, for spinning and looming into homespun cloth, in garden-sized plots. However, they could not grow cotton as a cash crop because of the labor-intensive process of hand cleaning the seeds from the fiber. There was no cheap labor in Jamestown as there was in India.

Meanwhile, in the early 1600s, Sea Island cotton was teamed with the churkha roller gin in the Caribbean and was much preferred over brittle green-seeded cotton from India by the emerging textile industry on both sides of the Atlantic. Long-staple black-seeded cotton became a cash crop in the Caribbean, but the more profitable sugar cane industry and rum trade usurped it. Around this same time, England began importing Indian cotton fabrics in the face of bitter opposition from British woolen interests. For some time in the early 1700s, the woolen interests of England succeeded in prohibiting the weaving and wearing of pure cotton fabrics. During the early part of the American colonial period, the fiber blend content of the clothing worn by the British and the colonists was typically 77 percent wool, 18 percent linen, and 5 percent cotton.[23] No wonder they were itchy people.

Fustian cloth, woven from a mixture of linen and wool, eventually gave way to flax and short-staple cotton. This cottage industry wove the weak Indian short-staple cotton threads only as the woof (the threads that run from edge to edge of the fabric) with the stronger flax threads as the warp (the threads that travel the length of the cloth). At that time, short-staple fibers could not be spun with a spinning wheel into thread strong enough to withstand the stress of the long warp threads on a loom.[24]

The early immigrants to the colonies came from rural England and had a lot of experience with spinning and weaving. They brought spinning wheels and hand looms and settled in New England to establish the colonial textile industry. Trade with the West Indies for long-staple cotton began in 1636. As spinning and weaving were home industries, nearly every home was a textile factory.[25] For example, in 1773, George Washington imported three bags of Indian green-seeded cotton to spin and weave fabrics for his family and the slave families on his tobacco plantation.[26] In support of this cottage industry, entrepreneurs on horseback stopped at homes throughout the country to supply raw materials and buy the finished thread and cloth.[27] Some of this cloth was sold in the neighborhood; some was exported overseas.[28]

In the 1760s, cotton planters in Virginia mounted a campaign to raise the income of their plantations by promoting finger-ginning and rod-ginning of green-seeded cotton. However, this effort was not continued, as they probably found that it was not economical to finger-gin or roller-gin green-seeded cotton. Nevertheless, in 1768, while Georgia shipped a total of 300 pounds of finger-ginned and rod-ginned green-seeded cotton, and South Carolina shipped 3,000 pounds, Virginia shipped a whopping 43,500 pounds.[29]

The packaging of cotton changed with the invention of the cotton baler, called the cotton compress, in about 1800. The enormous wooden screw was fashioned from an oak tree. The two legs, attached to the screw, were pulled around the baler by men or mules.

The Mechanization of the Weaving Industry

During the 1700s, a series of English inventions emerged that lowered the cost and increased the quality and availability of textiles.[30] These included the following:

- In 1733, John Kay installed a fabric, attached to a rope, across the vertical openings of the shuttle nests located on either side of a loom. Pulling on these ropes drove the shuttle back and forth, from side to side of the loom. This "flying shuttle" promoted faster weaving.[31]
- In 1761, James Hargreaves, a weaver and carpenter, carelessly knocked over a spinning wheel. He noticed

Gin and cotton compress

An early drawing of a cotton gin house and cotton compress screw.
*General Research & Reference Division, Schomburg Center for Research
in Black Culture, The New York Public Library, Astor, Lenox, and
Tilden Foundations*

that the now-upright spindle continued to rotate as
the large wheel rotated and spun thread. Three years
later, in 1764, he perfected the Spinning Jenny, which
he named for his daughter. It consisted of eight upright
spindles belted to a foot-powered pulley. He patented
the Jenny in 1770 and later equipped it with sixteen
spindles, which further improved the productivity and
reduced the labor cost. Children could, and did, operate
Spinning Jennies.[32]

- In 1769, Richard Arkwright harnessed waterpower to
  the Spinning Jenny and called the result a water frame.
  He freely borrowed ideas from earlier inventors and

Lincoln Spinning Jenny

James Hargreaves, a weaver and carpenter, invented the Spinning Jenny in 1764. It could spin up to sixteen spindles of thread at once. This Spinning Jenny is in the childhood cabin home of Abraham Lincoln and was owned by his mother, Nancy. *Library of Congress*

combined these features into a bank of spindles that drew out yarn and applied a twist to the thread. While many other textile inventors died in poverty, Arkwright became a multimillionaire, was knighted, and came to be known as the "father of the Industrial Revolution."[33]

- In 1775, Edmund Cartwright's steam-powered looms increased production.[34] Mechanization of the cotton textile industry began to strain the supply of cotton lint for the whirling spindles.

- In 1779, Reverend Samuel Compton's spinning mule combined Hargreaves' Spinning Jenny with Arkwright's water frame. This self-actuating hybrid spinner simultaneously drew out, twisted, and wound the cotton yarn into fine threads and reduced breakage. The spinning mule made stronger and finer threads in a range of sizes.[35]

The English textile technology ushered in the Industrial Revolution. These machines employed Thomas Newcomen's atmospheric engine, invented in 1712 and improved by James Watts in 1769. The mechanized weaving of cloth eventually destroyed the cottage weaving industries in England and North America.[36] In addition, due to these faster, more efficient machines, long-staple lint exports to England from the New World began to exceed short-staple exports from Asia and the Mediterranean.[37] Starting in 1786, the New World shipped about forty thousand pounds of long-staple lint to England every year.[38] Later, as American green-seeded cotton lint became available on the Liverpool market, weavers prized it over the brittle Indian green-seeded cotton.

In January 1783, fourteen-year-old Samuel Slater apprenticed himself for six and a half years to financier Jeremiah Strutt, a knitting mill manufacturer and business partner of Sir Richard Arkwright. During his apprenticeship in England, Slater learned how to design, service, and manage textile factories. He read an advertisement from America for skilled fabricators of textile machinery. In 1789, after completing his apprenticeship, he disguised himself as a farm boy, traveled to London, and slipped illegally out of England to sail for America. At the time, the law forbade skilled textile workers from leaving England because the English did not want these skilled workers to establish competing textile factories in other countries. Slater worked in New York for a year before he connected with a company needing his training and special talents.

On December 20, 1790, Slater became a full partner in the newly established firm of Almy, Brown, and Slater in Pawtucket, Rhode Island. While maintaining good relations with his original partners, Slater hunted for new mill sites throughout New England. Slater's strategy was to establish mills in new towns upstream from the old towns and to attract female employees from farm families. This strategy ensured a steady supply of both power and labor for the mill. With a new set of partners each time, Slater established a total of thirteen cotton mills on New England rivers, the last one in 1827 in South Oxford, Massachusetts. President Andrew Jackson later described Slater as the "father of American manufacturers."

Cotton 101

Seed cotton locks were picked from bolls of cotton plants by hand or by machine. These raw cotton fibers are interlaced with cotton seeds, which cling tenaciously to the fibers and must be removed before the cotton can be used.

The process of separating the seeds from the fibers is called *ginning*, and the device that does this is called a *gin*. The word *gin* is a shortening of the word *engine*. In the eighteenth century, an engine was any machine or mechanical device.

After they have been cleaned of seeds, the cotton fibers are called *lint*. Cotton lint is quality graded by its staple length, strength, diameter, foreign matter, and other factors. This lint is used to spin cotton thread and is woven into cotton cloth. The cotton seeds, separated from the lint in the ginning process, are planted to grow the next crop of cotton or are crushed, the vegetable oil expelled, and cottonseed meal is fed to cattle.

CHAPTER 2

Kincaid and Holmes

As early as 1784, South Carolina cotton farmers began shipping ginned cotton from Charleston to Liverpool.[39] One of the plantations the cotton came from was owned by Captain James Kincaid, who was born near Belfast, Ireland, in 1754 to Scottish parents and emigrated to America when he was nineteen. After Kincaid married Mary McMorries a few months later, they began living on Mary's dowry (a tract of land of unknown size) in what is now Fairfield County in South Carolina.

The Kincaids were a wealthy couple (Captain James eventually owned many ships), and in 1774, they began building a large, elegant, Georgian blue-granite mansion. The granite was mined from the famous ten-acre open-face quarry on the plantation, and the mansion stands to this day. Local tradition says that British General Cornwallis used the Kincaid mansion as his headquarters during his failed southern campaign in the latter years of the Revolutionary War.

If so, then the Kincaid family must have resided elsewhere because, in 1781, Captain Kincaid commanded colonial cavalry troops under Commanding General Nathanael Greene and Generals Francis "Swamp Fox" Marion, Daniel Morgan, and

Thomas Sumter. Some historians have declared Captain Kincaid the hero of the Battle of Eutaw Springs. Greene's army drove Cornwallis out of the South and ultimately to Yorktown, Virginia, where he surrendered to the American and French armies.

Kincaid was elected to the South Carolina House of Representatives in 1794 and served three terms. He died of yellow fever in Charleston, South Carolina, on October 20, 1801.

The production of green-seeded cotton on the Kincaid plantation is detailed in original family wills, probate records, abstracts of deeds, and accounting records for the mid-1790s. These records include the sale of cotton and purchase of farm products, as well as purchases of land, slaves, and foodstuffs. In particular, the Kincaid plantation made huge quantities of money from selling green-seeded cotton lint on the world market.

Ceo Forest, the factor (a middle man or commission man), maintained the ledger for the plantation. The ledger and other documents show that James Kincaid conducted an extensive business in buying, selling, and ginning cotton and operating a plantation store for 1795, 1796, and 1797. The table following provides an excerpt from the ledger, showing that the Kincaid plantation sold five bags of ginned cotton on May 23, 1795. This cotton would have been grown in 1794 and ginned during the winter and spring of 1795. The Kincaid plantation sold another fifty-eight bags (8,710 pounds) of cotton in 1795, and another twenty-three bags of cotton, weighing just over 4,800 pounds, were ginned in the fall of 1795. The ginning rate was 240 pounds per day.

In 1796, cotton output more than doubled to 124 bags (33,749 pounds). If the 1796 crop had been finger-ginned, it would have taken 149 slave families one year to gin that many bags of cotton.

These records of the sale of cotton from the Kincaid plantation suggest that at least two practical cotton gins were operating on the Kincaid plantation.

Entries from James Kincaid's Account Ledger Showing Sales of Cotton in 1795–1797

Date	Item	Quantity	Price (£.s.d)
1795 (cotton sold on May 23 was grown in 1794)			
May 23	5 bags of cotton	--	102.06.04
Nov. 18	30 bags of cotton	3,901 lbs.	231.01.02
Nov. 20	18 bags of cotton	3,764 lbs.	--
	5 bags of cotton	1,045 lbs.	--
		4,809 lbs.	280.10.06
1796			
Feb. 29	9 bags of cotton	2,349 lbs.	37.00.06
Nov. 1	82 bags of cotton	20,316 lbs.	--
	Divided with Col. Wm. Whitfield	7,676 lbs.	--
		12,640 lbs.	--

Date	Item	Quantity	Price (£.s.d)
Dec. 10	8 bags of cotton	2,226 lbs.	--
1797			
Feb.	7 bags of cotton	1,622 lbs.	
Mar. 10	8 bags of cotton	1,769 lbs.	
		18,257 lbs.	

We do hereby certify that the above mentioned three notes reg one for:
£149.4.11

£235.0.0

£235.9.9
Have been received by us and paid by
Mr. Ceo Forrest when due.
Signed Whistield and Brown
November 12th, 1799.

An analysis of the Abstracts of Deeds shows that James Kincaid, in addition to buying seed cotton and selling ginned cotton, was also buying land and slaves in the years that he was an early adopter of cotton saw gins and while serving in the South Carolina legislature. As for buying cotton, Kincaid would have used the profits from his own cotton operation to buy more from neighboring plantations. He may have begun growing, purchasing, and ginning cotton as soon as saw gins became available just after the Revolutionary War.

James Kincaid began buying land in 1788, according to Fairfield County probate records. The following table lists some of the land, slaves, and other property that he purchased.

Summary of Property Purchases
by the Kincaid Plantation

Year	Seller	Property	Book	Page	Price
1788	John Johnson	194 acres of land	C	198	100 pounds Sterling
1789	William Rabb	100 acres of land	A	298	100 pounds Sterling
1789	Henry Crumpton	40 acres of land	P	38	20 pounds Sterling
1789	William Starke	176 acres of land	C	195	100 pounds Sterling
1795	Christopher Ederington	Enslaved man	I	286	20 pounds Sterling
1795	George Ederington	Enslaved man	K	120	45 pounds Sterling
1796	Joseph Brevard	822 acres	K	214	1,500 pounds Sterling
1796	Thomas Roussam	One bay horse	K	292	6 pounds Sterling

Year	Seller	Property	Book	Page	Price
1797	Shadrick Jacobs	180 acres of land	L	162	16 pounds Sterling
1797	William Hardage	Enslaved male	L	64	100 pounds Sterling
1798	Enoch James	Horse and livestock	--	--	16 pounds Sterling
1799	Joseph McMorris	100 acres of land	M	39	50 pounds Sterling
1800	Hugh Milling	1,060 acres of land	N	36	620 dollars
1800	John Land	50 acres of land	N	46	50 dollars
Sum		2,722 acres of land			

While the Abstracts of Deeds listed Kincaid real estate purchases beginning in 1788, it is not known when James Kincaid actually began buying land. South Carolina began publishing the Abstracts of Deeds for Fairfield County in 1787, so there are no land sale data prior to 1787. Yet it is reasonable to believe that Kincaid was growing green-seeded cotton years before he began buying land.

Kincaid also apparently encouraged neighboring landowners to plant cotton, which he purchased and ginned along with the cotton from his own fields. This suggests that Hodgen Holmes had built and installed saw gins at the Kincaid mill and was ginning cotton in 1788, just as South Carolina folklore says. We believe that Holmes was actually ginning cotton for Kincaid several years before then.

Kincaid's two-story gristmill was built on the banks of Mill Creek. The gristmill eventually housed the cotton ginning operation, both driven by waterpower from the creek. In the ginning operation, large wool marketing bags were hung through holes in the second floor of the mill. A worker, standing in the bag, compacted the cotton with his feet while another worker kept the outside of the bag wet with water.

Sadly, the wooden buildings of the Kincaid plantation were burned during Sherman's March to the Sea in 1864. We believe an excavation of the site would discover important artifacts, including the circular saw blades and other fittings of the cotton gins and gristmills.

The will of James Kincaid, probated in 1810, called him a planter, storekeeper, and merchant. The will includes a list of the moveable inventory and the appraised value of the Kincaid plantation at Kincaid's death. In particular, the "appraisement of the estate of James Kincaid," dated August 12, 1802, lists a portable cotton gin with an appraised value of $120. This may be the small gin that Hodgen Holmes had fabricated in Hamburg, South Carolina, and later transported to the Kincaid plantation. The larger gins would have been permanently installed in the mill and were not considered moveable assets. Therefore, they are not listed in the inventory.

The inventory also lists male slaves at an average value of $452 and a female slave with two children valued at $770. One mule was inventoried at $75. Kincaid also owned land in Mississippi that was not included in the 1802 property inventory.

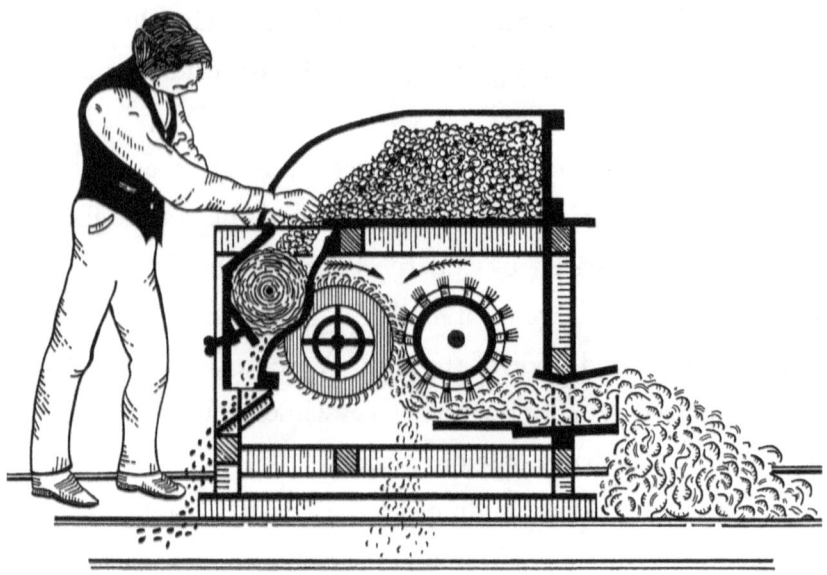

In about 1900, Daniel Tompkins, professor of mechanical engineering at Clemson University, reconstructed Holmes' saw-toothed cotton gin. *Lorraine Kuehnel*

Hodgen Holmes in South Carolina Folklore

South Carolina folklore states that, in the early 1780s, a local blacksmith, Hodgen Holmes, designed and built a rip-saw-toothed cotton gin. This early inventor may have heard about Walter Taylor of Southampton, England, who used a rotating circular saw to manufacture rigging blocks for the British Navy.[40] In any case, Holmes realized that circular saw blades could pull cotton fibers through a set of stripper ribs to clean the raw cotton of seeds. While South Carolina folklore states Holmes was the inventor of this cotton gin, Georgia folklore, like mainstream U.S. history, attributes the invention to Eli Whitney.

Interestingly, some versions of South Carolinian folklore say that James Kincaid was the inventor of the cotton gin. Kincaid may have seen a similar device during sea voyages: A saw blade rotating between wooden slats was used for cutting coconut husks without injuring the nut. Other folklore states that Kincaid may have told Holmes about Walter Taylor inventing the circular saw in England. A neighbor of Kincaid, William Watts, later testified that they tried to debur cockleburs from sheep wool with a rotating saw operating between slats.[41]

South Carolina folklore also states that Kincaid helped Holmes write an application for a caveat of invention, which, like a patent, prevented unauthorized use of an invention for a period of five years.[42] The U.S. Patent Committee, composed of President George Washington and Thomas Jefferson, may have granted a caveat of invention to Hodgen Holmes on March 14, 1789.

Little is known of Hodgen Holmes' early life. South Carolina folklore says that he was a carpenter and a blacksmith either in Hamburg, South Carolina (an incorporated town just across the Savannah River from Augusta, Georgia), or in Bamberg, South Carolina (in the southeastern part of the state). Both towns were located in green-seeded cotton growing country. Several attempts have been made to reconstruct the history of young Hodgen Holmes. These efforts have been published in a series of booklets, titled *Fairfield Sketchbooks*, copies of which are in the Winnsboro (South Carolina) city library. These sketchbooks provide credible information.

Advertisements recently gleaned from southern newspapers[43] published around the time of Holmes' death listed he had cotton gins for sale. Newspaper death notices for Holmes and notice of sales for disposing of his property after his death show that he manufactured cotton gins, wagons, and riding chairs. They also show he was a carpenter and blacksmith, had accumulated property, and owned slaves.

Holmes patent

Henry Ogden Holmes' patent for his rip-saw-toothed cotton gin in 1796. This was granted after Holmes received a caveat of invention for his cotton gin in 1789. *Courtesy of Robin Kaller of Kaller Historical Documents, Inc.*

Holmes' name appears on an 1805 list of those who invented "any new and useful art, machine, manufacture or composition of matter."
U.S. Patent Office

U.S. Patent Office

LETTER

FROM THE

SECRETARY OF STATE

ACCOMPANIED WITH

A LIST OF THE NAMES OF PERSONS

WHO HAVE INVENTED

ANY NEW AND USEFUL

ART, MACHINE, MANUFACTURE

OR

COMPOSITION OF MATTER,

OR

ANY IMPROVEMENT THEREON,

AND TO WHOM

PATENTS HAVE ISSUED

FOR

THE SAME,

FROM THE OFFICE OF THE

DEPARTMENT OF STATE,

With the dates and general objects of such patents, in pursuance of a resolution of the House of the 23d ultimo.

22d FEBRUARY, 1805.

Read and ordered to lie on the table.

AM. PHOTO-LITHOGRAPHIC CO. N.Y. (OSBORNE'S PROCESS.)

DEPARTMENT OF STATE,

February 18, 1805.

SIR,

IN pursuance of the resolution of the House of Representatives of the United States, dated the 23d of January last, I have the honor of transmitting a list of the names of persons, who have invented any new and useful art, machine, manufacture or composition of matter, or any improvement thereon, and to whom patents have issued for the same from this office, with the dates and general objects of such patents.

I have the honor to be,

SIR,

With the greatest respect,

Your obedient servant, &c,

JAMES MADISON.

To THE HON.
*The SPEAKER of the House of
Representatives of the United States.*

Subjects of the Patents.	Names of the Patentees.	Dates of the Patents.
Impt. in manufacturing sumach,	Joseph Hillard,	May 12th, 1796.
Improvement in bolting-cloths,	Robert Dawson,	May 12th, 1796.
Improvement in the cotton gin,	Hodgen Holmes,	May 12th, 1796.
Cleaning clover and other seeds, &c.	Jonathan Roberts, jun.	February 13th, 1796.
Improvement in piano fortes,	James Sylvanus M'Lean,	May 27th, 1796.
Improvement in burr mill stones,	Oliver Evans,	May 28th, 1796.
New invented steam engine and boiler,	Elijah Backus,	May 31st, 1796.
Stays for removing distortions in the spine,	Lunden M'Kechnie,	July 1st, 1796.
Improvement in sawing and polishing marble, &c.	Joseph Francis Mangin,	July 2d, 1796.
Machine for scouring rice and other grain,	Robert Grant,	October 17th, 1796.
Improvement in making salt,	George James,	November 16th, 1796.
Improvement in manufacturing pot ash,	Edward Ryan,	November 16th, 1796.
Improvement in concentrating the volatile parts of calcareous earth, stones, &c.	John Fowler,	November 16th, 1796.
Improvement in manufacturing cut nails,	Peter Cliff,	November 16th, 1796.
Machine for heading and cutting nails,	Isaac Garretson,	November 16th, 1796.
Improvement in a printing press,	Apollos Kinsley,	November 16th, 1796.

U.S. Patent Office

Subjects of the Patents.	Names of the Patentees.	Dates of the Patents.
Improvement in splitting sheep skins,	James Stansfield,	November 16th, 1796.
New method of ruling books and paper,	Mark Isambard Brunel,	November 16th, 1796.
Improvement in pumps,	Theobald Bourke,	November 16th, 1796.
Manufacturing pot ash,	William Frobisher,	November 17th, 1796.
Improvement in forging bolts and round iron,	Clement Rentgin,	November 17th, 1796.
Improvement in a loom for weaving cloth,	Amos Whittemore,	November 17th, 1796.
A "preambulator" for measuring a ship's way,	Amos Whittemore,	November 19th, 1796.
Machine for cutting nails,	Amos Whittemore,	November 19th, 1796.
Improvement in manufacturing nails,	John Bigelow,	November 19th, 1796.
Improvement in cutting and heading nails, &c.	George Chandlee,	December 12th, 1796.
Conjurer for cooking and boiling,	Thomas Passmore,	December 23d, 1796.
Improvement in manufacturing cut nails,	Daniel French,	December 23d, 1796.
Improvement in manufacturing wrought nails,	Daniel French,	December 23d, 1796.
Improvement in heading nails,	Jared Byington,	December 23d, 1796.

U.S. Patent Office

Holmes may have discussed cotton ginning with tobacco and indigo planters in the early 1780s. He may have learned about and even built churkha gins and ginned black-seeded cotton purchased from the tidewater area of Georgia. He probably tried, and failed, to gin green-seeded cotton with churkha and barrel gins.

Regardless of the source of Holmes' inspiration for his cotton gin, South Carolina folklore says that Kincaid learned that Holmes had invented and fabricated a practical cotton gin, and Kincaid asked Holmes to bring his new gin to the Kincaid plantation. In 1787, South Carolina folklore states that Holmes transported his original eight-bladed cotton gin from Hamburg, South Carolina, to the Kincaid plantation near Winnsboro, South Carolina, where lint quality compared favorably with finger-ginning.[44] Holmes later installed larger gins in Kincaid's water-powered gristmill on Mill Creek[45], for ginning the green-seeded cotton that Kincaid grew and that he purchased from his neighbors, again according to local folklore.

Holmes' practical, continuous-flow cotton gin was equipped with eight saw blades. He had the blacksmithing ability and raw materials necessary to fabricate the circular saw blades, as well as for making shafts, handles, and other metal parts. While no Holmes-made gins have ever been found, they were offered for sale by advertisements in the *Augusta Chronicle* dated September 15, 1804.[46]

An Analysis of Holmes' Cotton Gin

Holmes incorporated the following elements into his cotton gin:

1. He cut saw teeth around the perimeter of eight- to fourteen-inch diameter iron or steel disks. He likely used old felling saw blades or other flat scrap metal as the raw material for the circular saw blades.

2. He made a seventy-two-inch-long cylinder by alternating saw blades and one-inch-thick wooden spacers on a wrought-iron shaft.[47] He fabricated bearings for each end of the shaft and attached a crank handle. The later, larger gins were driven by waterpower rather than hand cranks.

3. The Holmes' patent specifications describe a feeder cylinder located in the hopper, in front of the saw blades. This feeder cylinder does not appear in any illustrations of the gin[48] and apparently was later eliminated.

4. He installed flat, vertical ribs with thin slots between the ribs. The saw blades protruded through these slots into the hopper, which held the raw cotton. The roll box (hopper) was located between the ribs and the front side of the gin box. The inner side of cotton mass was continuously powered and rotated upward by the rotation of the saw blades.

5. As the saw blades rotated, the teeth snagged fibers from the rotating cotton mass. The fibers were pulled through the slots by the saw teeth, as the attached seeds collided with the ribs. Because the slots were narrower than the seeds, the seeds were stripped from the fibers, creating clean cotton lint.

6. A very important feature of Holmes' gin was the geometry of the roll box (hopper) that allowed the seed to fall out by gravity. Whitney's gin required that the ginning operation be stopped and all cotton, ginned and unginned, be removed by hand. Evidently, Judge William Stephens did not appreciate the major difference in the two gins.

7. On the far side of the saw blade cylinder, behind the ribs, a cylindrical brush rotated at about five times the speed of the saw blades, and in the opposite direction. This doffer brush was fitted with hog bristles that stripped the cotton lint downward from the saw teeth as fast as the fibers were collected through the ribs by the saw teeth.

Holmes described his gin in his patent application.[49] However, he submitted no drawings and mentioned rip-saw teeth only once in the specifications. In 1844, in support of Holmes' claim to the rip-saw-toothed design, William Benjamin Seabrook wrote: "The Holmes machine was set up in the grist mill of Capt. James Kincaid on Mill Creek in Carven (now Fairfield) County, South Carolina, in 1795, and it is reported to have been the *first rip-saw-toothed* gin [italics added] used in the state."[50] It was powered by a water wheel.

Since Holmes' caveat of invention would have had a life of five years, it would have expired on March 14, 1794. Unfortunately for Holmes, the U.S. Patent Office failed to automatically upgrade his War Office's caveat of invention to a patent, which would have had an additional life of nine more years. Dr. M. C. McMillan, author of *The Manufacture of the Cotton Gin*, states that the day Holmes' caveat of invention for his saw gin expired was the same day that Whitney's bent-needle gin was patented: March 14, 1794.

Documents Concerning Holmes' Gin

In an interesting letter dated June 13, 1937, W. H. Hutchinson, secretary treasurer of the Cotton Seed Crushers' Association of

Georgia, asked Janie Hutchinson of Monticello, South Carolina, a question nearly all South Carolinians ask, even to this day. He wanted to set the record straight as to who invented the rip-saw-toothed cotton gin. W. H. Hutchinson wrote as follows:[51]

The invention of the cotton gin is given as the second most important invention in all human history: first place is accorded to the steam engine.

With the lights before me, I am disposed to give the cotton gin first place, because it unquestionably revolutionized world commerce and influenced civilization.

Now, the foregoing being true, is it now incumbent on us to establish and proclaim to the world the true history of the invention of the cotton gin? I think it is.

All published history accords to Eli Whitney, the Yankee school teacher, the honor of inventing the cotton gin. I decline to accept this.

Eli Whitney was unquestionably intelligent. Being a graduate of Yale College and having been reared in New England, where the history and the school books of the time were published, he was necessarily intelligent. He came south and entered an atmosphere wherein a most serious problem was the all-absorbing topic—a means of separating the fiber of the cotton from the seed.

The local people were not asleep. I am convinced that a great deal of preliminary inventive progress had been made when Whitney arrived. Joseph Watkins, of Petersburg, Ga., made claim prior to Whitney. His descendants give strong assertions that Whitney visited him and appropriated his work.

Cotton Gin 1793

An early cotton gin in operation in 1793. *General Research & Reference Division, Schomburg Center for Research in Black Culture, The New York Public Library, Astor, Lenox, and Tilden Foundations*

Whitney's visit to the Kincaid place, where Hodgen Holmes and Mr. Kincaid were working on a machine, leads me to the conclusion that having plenty of money, which Mrs. Nathanael Greene supplied, and an abundance of Yankee shrewdness, he appropriated everything he could get, either by fair or foul means, and rushed to Washington for a patent!

However, he secured a patent only for a spike gin, a thoroughly impractical machine. I have a copy of his patent. It was Hodgen Holmes who invented the [circular saw] gin. I have a copy of his patent.

Finding old cotton gins

In 2006, Dr. Buchele went to South Carolina to do further research for this book and look for old cotton gins in Fairfield County. Here are two he discovered in old barns. *Above: Fairfield County Historical Society; Below: Bishopville Museum*

To realize that so little is known concerning Holmes and his family is most disappointing to me, and I am anxious to locate the greatest event in history in your community—in fact, near Mr. Daniel Heyward's hospitable home [Heyward was the current resident of the Kincaid mansion].

Is it not possible to "set the wheels in motion" that will accomplish this? It does seem to me we should be able to locate some of the descendants of Hodgen Holmes who could give his life history, when and where he was born, when he died, where buried, etc. Maybe he left something written about his gin. I have found that sometimes the genealogy of families comes to light unexpectedly.

I may say that [I] have completed a manuscript of "The Story of the Cotton Seed Crushing Industry" except the important feature the "The Story of the Cotton Gin." I am withholding my story from the press in hopes I may develop and include the true story of the gin.

I am sending a copy of this letter to Messrs. Daniel Heyward and S. T. Dunn, and hope our combined efforts will develop something.

I shall add to my story reference to the Old Brick Church—the inscription on the wall by the "dam Yankee," Mr. Heyward's wonderful restoration of the old home, etc.

Sincerely yours,

Signed

W. H. Hutchinson

Secretary Treasurer

Mr. Hutchinson's letter is a grand introduction to the question of who really invented the cotton gin. This question arose shortly after the Revolutionary War. The question still existed in 1937, and it exists to this day.

Holmes manufactured gins in Augusta, Georgia, until his death on December 28, 1804. It is not known when he began his gins. However, Judge Williams Stephens nullified his patent on November 6, 1802. Therefore, in March 1803, Holmes purchased a license to manufacture rip-saw-toothed gins from Eli Whitney.

As late as December 7, 1802, Holmes still advertised that "he is the only person with authority to sell cotton gins equipped with metallic plates, commonly called the rip-saw-toothed gins." Holmes must have enjoyed tweaking Judge Stephens and Whitney's ego in this public announcement. Holmes' wife, Elizabeth, later advertised the sale of three saw gins, and Holmes' tools and equipment, by auction and published his obituary in the *Augusta Chronicle*.[52]

On November 16, 1802, *The South Carolina State Gazette and Columbia Advertiser* printed the following advertisement:

"CAUTION"!

THE public are hereby notified, that the Subscriber has the only Patent which has ever been issued under the authority of the United States, for the COTTON gin worked with metallic Plates, commonly called the Rip-saw-toothed, and consequently that no other person has a right to sell or use that invention but himself, or such persons as are duly authorized by him.

HODGEN HOLMES.

The Printers throughout the Southern States are requested to insert the above.

On December 7, 1802, Holmes advertised again in *The South Carolina State Gazette and Columbia Advertiser* that he was the only person with authority to sell the cotton gin equipped with "metallic Plates, commonly called the Rip-saw-toothed gins." On September 15, 1804, *The Augusta Chronicle* printed the following advertisement:

FOR SALE, TWO COTTON GINS.
Reckoned to be as good as any in this part of the country, will be sold low for cash, or for seed cotton. Also two RIDING CHAIRS, with harness to one complete, which will be sold for low ready money.
HODGEN HOLMES.

On December 1, 1804, *The Augusta Chronicle* printed a second, similar advertisement:

FOR SALE, TWO NEW COTTON GINS.
Reckoned to be as good as any in this part of the country, will be sold low for cash, ALSO, Two new RIDING CHAIRS, with harness to one complete, and a fishing SEINE, partly new, in use only one season.
HODGEN HOLMES.

On December 28, 1804, *The Augusta Chronicle* printed the following obituary:

DIED.
On Friday, the 28th, after a long illness, which he bore with great fortitude and patience, Mr. Hodgen Holmes, a respectable citizen of Georgia and a useful member of society.

On January 12, 1805, *The Augusta Chronicle* printed the following three advertisements regarding the disposal of Holmes' estate:

WILL BE RENTED.

At the Market Square in the city of Augusta, on Saturday next, the 19th inst. A LOT containing 50 acres, belonging to the estate of Hodgen Holmes, deceased.

At the same time and place, will be hired the Negroes belonging to the said estate.

ALSO A FISHING SEINE, to be sold.

Terms made known on the day of the sale.

ELIZABETH HOLMES ADM'X.

WILL BE SOLD.

On the 23rd day of February- next, personal estate of Hodgen Holmes, dec. Consisting of horses, cattle, wagons, carriages, a set of blacksmith tools, Plantation and Chair maker's tools, three Cotton Gins, household and kitchen furniture, with a variety of other articles.

Conditions

All sums not exceeding Twenty Dollars, cash. . . . And all over that, twelve months credit with approved security.

ELIZABETH HOLMES ADM'X

NOTICE.

All persons having demands against the estate of Hodgen Holmes, dec. are requested to come forward as the law directs, and all those indebted to the said estate, are also requested to make immediate payments to,

ELIZABETH HOLMES ADM'X.

Holmes may have become wealthy manufacturing gins as well as riding chairs, carriages, and wagons. The February 23, 1805, sale inventory showed that Holmes developed a manufacturing business in the estimated eight years (1796 to 1804) of manufacturing cotton gins. The date he left the Kincaid plantation and began manufacturing gins is not known. Our best estimation is that he began manufacturing gins the day he stopped operating the Kincaid gins.

Holmes advertised three gins for sale between September 15, 1804, and December 1, 1804. The list of tools for making riding chairs and the listing of two riding chairs for sale indicate that he was also in the carriage business. He owned a fifty-acre farm, livestock, and the tools of the trade for carpentry and blacksmithing.

In the end, though, Holmes may not have been the first to conceive and fabricate a rip-saw-toothed cotton gin. South Carolina folklore states that he was the first to prepare an application for a caveat of invention for a practical, continuous-flow, rip-saw-toothed cotton gin; the first to file a caveat of invention for a saw-toothed cotton gin to the U.S. War Office; the first to be granted a caveat of invention for a saw-toothed cotton gin; the first to file for a patent application for a continuous-flow, rip-saw-toothed cotton gin; and the first to be granted such a patent by the U.S. Patent Office. In addition, cotton gins based on Holmes' practical design, not Whitney's, have ginned nearly all the cotton grown in the world since the 1780s.

CHAPTER 3

Whitney and Miller

Southern folklore and historians have connected Hodgen Holmes and Eli Whitney, both inventors of cotton gins. Some historians believed Whitney worked for Holmes at the Kincaid plantation. Others claimed Whitney worked in Holmes' blacksmith shop in Hamburg, South Carolina.[53]

Mr. Anderson, a descendent of Captain James Kincaid, wrote the following about Whitney's connection to Holmes and his great-grandfather:

Still other Fairfield historians state that while Kincaid and Holmes were in Charleston, S.C. on cotton business in winter of 1792–1793, a young man requested permission of Mrs. Kincaid to examine the gristmill. He represented himself as a school teacher in Winnsboro, South Carolina. She gave him a key; he examined the mill and the gins, returned the key, and rode off. Mr. Kincaid later learned that the young man was Eli Whitney, a school teacher from Savannah, Georgia. The same story is told about Whitney visiting Petersburg, Georgia, and trying to steal Joseph Watkins' roller-gin invention that Joseph Watkins would patent on December 23, 1796.[54]

After learning of Holmes' gin from Eli Whitney, Phineas Miller, a lawyer friend who was aware of the value of patent filing dates, urged Whitney to begin fabricating a cotton gin of his own and writing a patent application. Whitney then traveled to Philadelphia to learn about filing a patent application from Thomas Jefferson.

Whitney Meets Miller

Eli Whitney was born in Westborough, Massachusetts, on December 8, 1765. He was a precocious farm boy. Too young to shoulder arms in the Continental Army, he manufactured square nails in his father's forge to support the war effort.[55] Whitney probably observed the design of his mother's bent-needle-toothed carding paddles and may have been aware of the bent-needle-toothed carding drums invented by Paul Lewis in 1748 and Daniel Bourn in the 1750s. He may have also learned of Oliver Evans' 1777 invention of a machine for making bent-needle, leather-backed carding cloth at the rate of three thousand bent carding needles per minute.[56] He also attended and graduated from Yale.

So did Phineas Miller, who was hired in 1785 to tutor Major General Nathanael and Catharine Greene's children at their Mulberry Grove plantation in Georgia and read law. In those days, a graduate of a liberal college became a lawyer by reading law under the tutorage of a lawyer. After General Greene died of sunstroke on June 19, 1786, the general's widow appointed Miller the manager of the plantation.

In the late summer of 1792, while Mrs. Greene and Miller were vacationing in Connecticut and Rhode Island, Miller met

Whitney (a fellow graduate of Yale College) and recruited him for a tutoring job with Major DuPont of South Carolina for the following winter. Whitney had just learned that a promised teaching job in New York City, where he intended to read law, had evaporated. Miller introduced Whitney to Mrs. Greene as they sailed south on a coastal packet from New York to Savannah, Georgia.

During the seven-day trip south, Whitney, Miller, and Mrs. Greene surely discussed the high prices that American-grown green-seeded cotton lint was commanding on the London market, the wealth that could be earned growing black-seeded cotton on the tidelands, and the apparently unmarketable green-seeded cotton grown in the region's vegetable gardens.[57] They undoubtedly discussed the high prices paid for long-staple cotton cleaned with imported churkha gins and the establishment of textile mills in New England that were equipped with the latest spinning and weaving machines.

During their leisure time on the boat, Miller and Whitney likely also read law and caught up on their correspondence. Because Whitney and his associates were prolific letter writers, and because they were later involved in a large number of patent infringement litigations that were recorded verbatim, most authors on the history of the cotton gin begin their stories with Whitney's arrival in Georgia, giving exact dates such as June 20, 1793. This is actually the date when Whitney met Thomas Jefferson in Philadelphia to discuss filing the patent application for his bent-needle cotton gin.

After arriving in the South, Whitney was invited to stay at the Mulberry Grove plantation until his teaching duties started. According to his sister Elizabeth's recollections in 1836, Whitney passed the time mending the children's toys, fixing Mrs. Greene's embroidery frame, and enjoying southern hospitality. Whitney then learned that his tutoring salary would be cut in half. He resigned the South Carolina job and accepted the job of tutoring Mrs. Greene's children while reading law under Miller's tutorship.

In the 1780s and 1790s, many residents in this region would have been engaging in long discussions about the merits of green-seeded versus black-seeded cotton, how to clean seeds from green-seeded cotton, and the design of a practical cotton gin. Whitney and Miller continued studying the problems of ginning cotton that they had discussed on the voyage down to Georgia. They probably purchased or built a churkha and ginned black-seeded cotton. They then tried ginning green-seeded cotton with a churkha and observed how the seeds were crushed as they passed through the pinch point of the rollers, soiling the cotton with oil. They may have tried modifying the rollers for ginning green-seeded cotton to no avail.

By coincidence, the two American cotton gin inventors were virtually neighbors. Hodgen Holmes lived in Hamburg, South Carolina, across the river from Augusta, Georgia, and about ninety miles up the river from Mulberry Grove plantation. The Mulberry Grove plantation was located in the black-seeded cotton area of the tidewater, fifteen miles upstream from Savannah, Georgia. Whitney, Miller, and Mrs. Greene may have learned about the rapid increase in green-seeded cotton production and the invention of a rudimentary rip-saw-toothed cotton gin by Hodgen Holmes in South Carolina. They also may have known about the increasing shipments of cotton from Charleston, beginning in 1784.

Whitney wrote to his family concerning the development of the cotton gin, but he professed on several occasions, in letters, that he never left the Mulberry Grove plantation except to purchase seed cotton from a warehouse in Savannah, Georgia. However, South Carolina folklore states that Whitney traveled widely in the green-seeded cotton country that was, at that time, in a cotton-ginning ferment. We believe he visited cotton gin inventors such as Holmes while he scouted out cotton gins, just as his sister Elizabeth stated: "Whitney traveled west in his youth to find an employee for his forge shop and to learn about blacksmithing." South Carolina folklore also states that Whitney traveled to Winnsboro, South Carolina; visited Holmes and studied Holmes' cotton gin; and may even have been employed in Holmes' blacksmith shop.

Several stories in Georgia folklore recall Whitney's inspiration to invent a cotton gin. The classic version, but by no means the only version, is as follows:

During the late winter of 1792–93, Mrs. Greene entertained Majors Brewer, Forsyth, and Pendleton, Revolutionary War Army comrades of her late husband, Major General Nathanael Greene. The major's plantations were located in the green-seeded cotton growing area around Augusta, Georgia. After dinner, their conversation quickly turned to the exciting problem of 'cleaning' the seeds out of green-seeded cotton now grown in their gardens and expanding production onto their plantations and their dream of instant wealth!

Catharine listened for a while and then made her famous assertion! 'Surely, Mr. Whitney can supply your equipment needs; he can make anything!'

While the Army Majors regarded her remarks as a pleasantry, Whitney took them seriously and began working on fulfilling the Southern planter's dreams of a simple, high capacity gin for separating seeds from green-seeded cotton.[58]

The fact that Whitney traveled into the green cotton country is reported in the following article from the *Columbia Register* about the 1881 New Orleans Exposition. This article also supports the South Carolina folklore about the invention of the practical, rip-saw-toothed cotton gin by Holmes in 1787:

New Orleans 1881 Exposition Exhibits the Rip-Saw-Toothed Cotton Gin

Among the South Carolina exhibits at New Orleans will be the original Letters Patent on parchment, signed by G. Washington, President, and granted to H. Holmes, of South Carolina, for a cotton gin. A letter accompanies the patent, written by Mr. Geo. H. McMaster, of Winnsboro, S.C., which expressed the belief that Whitney filched the invention from Holmes, and adds, James Kincaid, a soldier of the Revolution, being told by his friend, Holmes, who lived near Hamburg, in this state that he had invented a cotton gin, agreed to take the gin and try it at his mill, which was located in the western part of Fairfield County. He did so, and while the mill was closed for a few hours, in the absence of Kincaid, a young man rode to the house and requested of Mrs. Kincaid permission to examine the mill. She, forgetting the injunction of her husband not to permit anyone to enter the mill during his absence, gave the key to the young man; he returned it in a short time and rode off.

Mr. Kincaid subsequently learned that the young man was Whitney, and this is believed by Kincaid's descendants, who still own the mill site. The old, original cotton gin burned along with the mill, at the time of Sherman's destructive march through the State. Dr. Wm. Cloud, who married a daughter of Holmes, preserved the parchments. Accepting it as true that the cotton gin was the invention of a South Carolinian, it will be seen that she has led all the States in everything connected with the great southern staple. She invented the cotton gin, and a legislature was the first to pay a royalty for its use. The only improvement on the rip-saw-toothed has recently been patented by a South Carolinian; the "Cotton Harvester" is a South Carolina invention.

Whitney wrote his father on September 11, 1793: "After seeing cotton for the first time, I have learned about the difficulty of cleaning seed from the fiber."

Whitney Builds a Cotton Gin

When Whitney returned from his trip to the upland green-seeded cotton-growing country of South Carolina, he told Miller about seeing rudimentary saw gins cleaning green-seeded cotton. Together, the lawyer and apprentice hatched a devious plan. They knew about the newly revised U.S. Patent Office, and they guessed that the South Carolina cotton gin fabricators were still ignorant of the existence of the patent office and its purpose. They decided that Whitney would build a cotton gin as fast as possible and then write a patent application for it. They would then use their senior patent and knowledge of the law to gain control over all cotton gins and all cotton gin patents, granted before or after their patent was granted, and thereby control the cotton ginning industry and its profits.

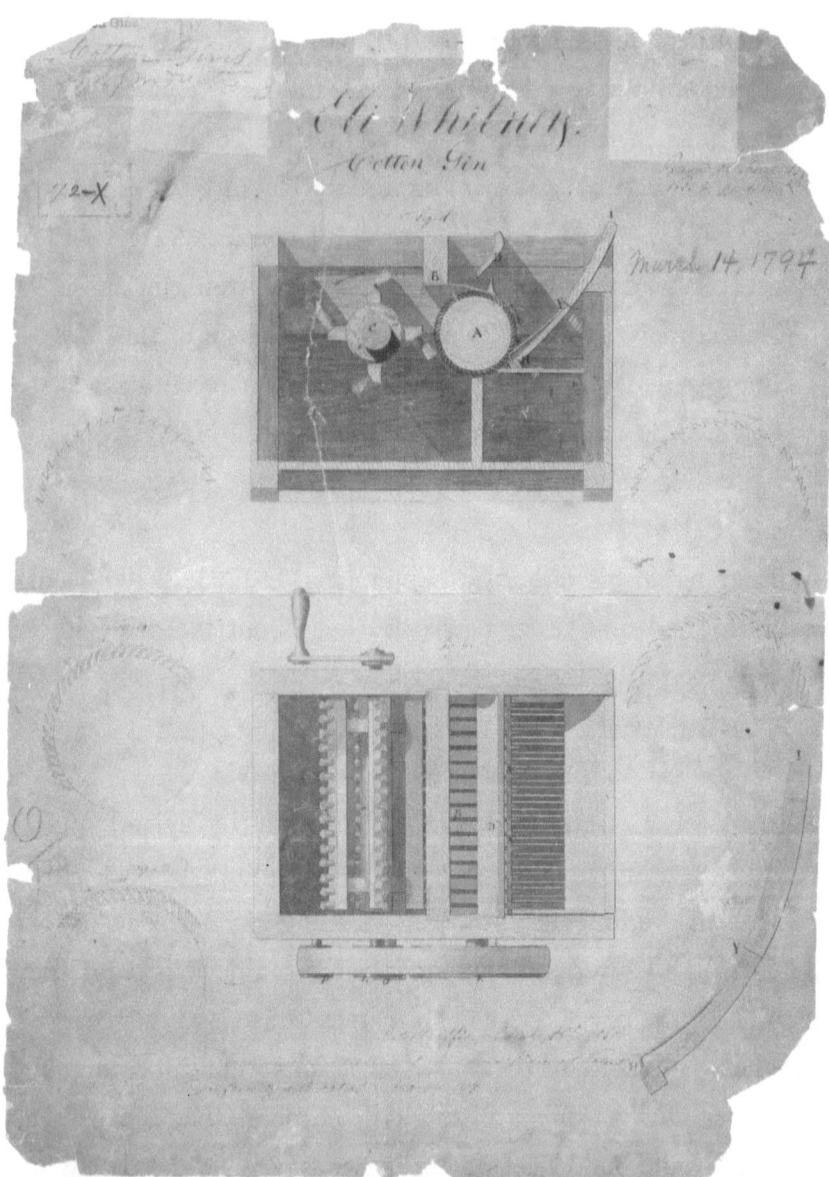

Patent for Whitney's cotton gin (1794)

Eli Whitney's original cotton gin patent was issued March 14, 1794.

National Archives

We believe that Whitney learned about ginning cotton during his trip through South Carolina, where he studied the saw-toothed ginning process. On the other hand, Georgia folklore speculates that Whitney got his idea for ginning cotton when he saw a cat clawing feathers from Mrs. Greene's canary by pawing through a wire birdcage.[59] This folklore states that Whitney tested his idea by pushing a hooked wire through a narrow slot into a basket of seed cotton. When he pulled the hooked wire out of the basket, the hook held only cleaned cotton. The seeds had been stripped off and remained behind in the basket. Whitney demonstrated this idea to Miller. After witnessing the demonstration, Miller, in his capacity as plantation manager, loaned Whitney a workroom in the basement of the house and offered to pay for all development expenses. In case of failure, Whitney would only be out his time. Whitney accepted Miller's offer.

Whitney did not set out to make a saw gin. He had no access to sheet metal or the means of fabricating it. Instead, we believe Whitney remembered his mothers' carding paddles, equipped with bent-needle pins, which hung by the hearth in Massachusetts. The bent-needle pins hooked wool fibers as the carding paddles were pulled in opposite directions, straightening and cleaning the fibers, which could then be drawn and spun into yarn. Whitney visualized rows of bent-needle teeth mounted in annular rows around a rotating cylinder, much like the carding drum invented by Paul Lewis in 1748 and Daniel Bourn in the 1750s. Each of the hundreds of needles had to be flattened on one end and driven into the wooden roller. Each needle then had to be bent, using pliers, to a precise angle and hand-sharpened with a file. To make the needles, Whitney drew down the diameter of No. 12 gauge steel wire (7/64-inch diameter) by pulling the wire through a die to create No. 14 wire (5/64-inch diameter). This process also work-hardened the wire, making it stiffer. Mrs. Greene purchased the wire in New York City.

Whitney then cut the drawn wire into one-inch lengths and flattened one end of each needle. While holding the needle with grooved pliers, he drove the needle into the wooden cylinder, with the flat end across the grain. He discovered that when he drove the flattened ends of the needle parallel with the grain, the wood tended to split.[60] He then bent the needle forward in the direction of travel, just like the pins on a carding paddle.[61]

The rows of bent needles were aligned with a slotted wooden breastwork.[62] As the cylinder rotated, the bent needles passed through the mass of cotton seed and snagged cotton fibers. The rotating cylinder carried the cotton seed with fiber onto the apron of the horizontal breastwork and then into slots sawed in the breastwork. The seeds then were stripped from the fibers at the entrance of the slots in the breastwork and were spilled back onto the apron of the breastwork and back into the hopper and rejoined the mass of raw cotton and ginned seeds. The more cotton that was ginned, the more seeds that were collected on and in front of the breastwork. This slowed the ginning process. When the hopper became full, the cylinder had to be stopped and the mixture of stripped seeds and raw cotton needed to be removed from the hopper by hand. This mixture batch then had to be sorted and the unginned seeds returned to the hopper for further ginning.

This was an important departure from Holmes' design. Whitney's gin used a different type of teeth (bent needles versus rip-saw teeth cut in a steel disk blade); slotted, horizontal breastwork versus stripping ribs; and a different-shaped hopper. Most importantly, Holmes' gin was an efficient, continuous-flow design. The rotating roll of raw cotton in Holmes' gin provided more opportunities for the saw teeth to hook onto fibers. In addition, the seed-stripping ribs in Holmes' design were vertical. As soon as the vertical ribs stripped the seeds, they fell back onto a rotating seed roll and were carried around to the screen openings in the bottom of the hopper. The ginned seeds fell through holes in the bottom of the roll box and onto the floor. The cleaned cotton seeds did not accumulate inside the hopper, so Holmes' gin was a continuously operated gin.

Whitney's gin was batch-type design. It had to be filled with raw cotton, operated, and then stopped and cleaned out by hand when the hopper became full of ginned and unginned cotton. This stop-and-go operation made Whitney's design impractical for large-scale operation, as in James Kincaid's water-powered mill. In addition, the mass of raw cotton did not rotate. It remained an unconsolidated mass on the apron of the horizontal breastwork, and the stripped cotton seeds accumulated on it. This was not all, though. The mass of ginned seeds that were removed from the box still contained unginned cotton. The mixture had to be sorted by hand and the unginned cotton returned to the roll box.

Early in 1793, Whitney wrote to Josiah Stebbins, complaining that "one of my best cylinders failed." He completed the working model of his cotton gin in late April 1793, in New Haven, Connecticut.[63] According to Georgia folklore, in the winter of 1792–1793, Mrs. Greene assembled a group of distinguished guests to see Whitney's work in progress. During a demonstration, after a few turns of the hand crank, the bent needles of the experimental gin each collected a wad of fibers. The cylinder quickly became covered with blanket of ginned cotton. The mass of fibers began to press up against the breastwork, raising the wooden ribs, and uncleaned seed cotton began to accumulate on the cylinder

Whitney became embarrassed; he feared he had failed. However, when Mrs. Greene saw the teeth of the gin clogged with cotton, she took a broom from the hearth and briskly swept the fibers downward off the rotating cylinder. The fibers fell off cleanly and accumulated on the floor. Inspired, Whitney immediately constructed a counter-rotating doffing cylinder with tufts of hog bristles that cleaned the bent-needle cylinder, exactly as Holmes had done at least six years earlier for his saw gin.

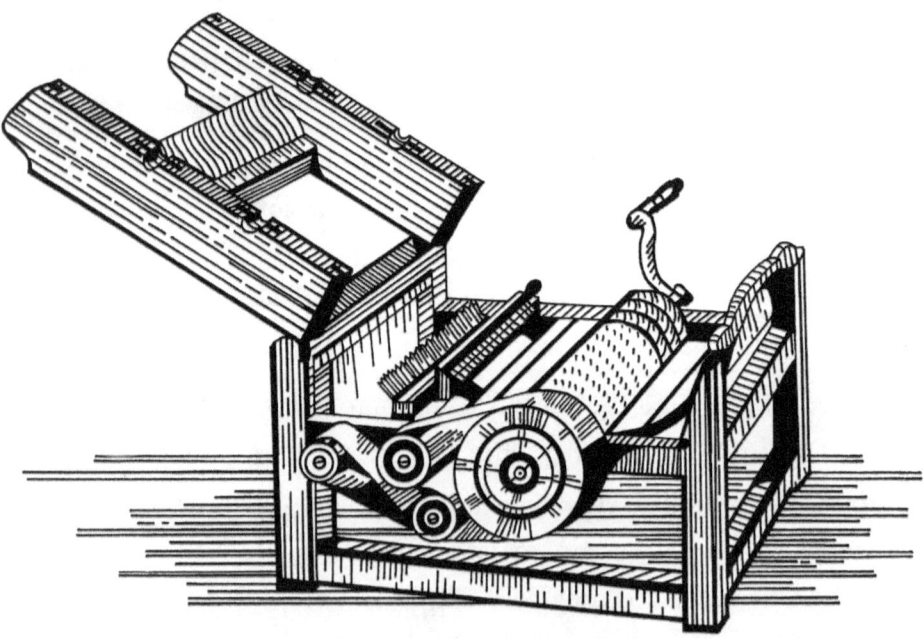

Whitney's original batch-type, bent-needle gin of 1793. The breastwork is hinged at the back. *Lorraine Kuehnel*

This doffer brush rotated five times faster than the bent-needle cylinder and efficiently brushed the fibers from the bent-needle teeth.[64]

Beyond that addition, Whitney did not try to improve his first gin. Instead, he patented the first design that actually worked. In the summer of 1794, Whitney began the laborious process of manufacturing gins with hand-cut and hand-bent needles. Whitney continued to manufacture this impractical design even after his partner, Miller, began substituting Holmes-like saw-toothed cylinders in Whitney's gins. Even then, Whitney continued to make batch-type gins. Whitney never understood how the design of the hopper allowed Holmes' gin to be a continuous-flow gin. And Whitney never realized that Holmes' saw gins could be mass produced and manufactured with interchangeable parts—until years later, after he began manufacturing muskets and had a supply of sheet steel available for making saw blades.

An Analysis of Whitney's Cotton Gin

Whitney started with a wooden cylinder about six inches in diameter. The center of this cylinder was bored out, and a wrought-iron axle was inserted and mounted on bearings. A crank handle was attached to one end of the axle. To create the annular rows of bent-needle teeth on the rotating wooden cylinder, Whitney flattened one end of each wire with a hammer. The flattened end reduced the tendency of the needle to twist sidewise while gathering fiber or pulling the fiber through the slot of the breastwork.

The horizontal wooden breastwork lay on top and tangent to the rotating cylinder. The breastwork was attached to a pair of rearward extending arms that were hinged to the back of the gin.

Raising the breastwork exposed the rotating cylinder. Brass or iron wear plates were attached to the underside of each rib of the breastwork. These wear plates lay on the rotating cylinder, and the chisel-like points of the wear plates were tangent to the rotating cylinder. The bent-needle teeth rotated through the narrow slots in the breastwork and carried the cotton fibers into the slots, stripping off the cotton seeds. The stripped cotton seeds rejoined the cotton mass that lay on the apron of the breastwork downstream from the hopper.

A rectangular box enclosed the cylinder, the hopper (roll box), the breastwork, and the bearings. The cotton gins invented by Holmes and Whitney both required workers to feed green-seeded cotton into the hoppers and workers to turn the crank that drove the cylinders. Both gins self-fed the cotton fibers into the rotating teeth. The two designs employed distinctly different designs for snagging the fibers (saw-blade teeth verses bent-needle teeth), but both cylinders rotated upward with respect to the mass of cotton in the hopper. Both gins were equipped with similar, downward-rotating doffer brushes for cleaning the snagged cotton off the slower, downward-rotating teeth of the cylinders.

Holmes' gin had a capacity of several hundred pounds of cotton per day, while Whitney's gin cleaned about one hundred pounds of cotton per day. Holmes' continuous-flow saw gin required no adjustment, just an occasional sharpening of the teeth. Whitney's bent-needle gin required constant sharpening, bending, straightening, or replacement of the bent-needle teeth.

A time and motion study revealed that an average of ten seconds was required to cut, flatten, and drive one needle into the roller and bend it forward. Whitney's patent specifications gave explicit instructions on how to install the needles and bend them.

It specified the distance (based on the width of a cotton seed) between the 5/64-inch diameter bent-needle teeth. The annular rows were 7/16 inch (the length of a cotton seed) apart. The length of the log was 48 inches and incorporated 88 annular rows of bent-needle teeth. Therefore, a 6-inch diameter cylinder required 11,611 needles: 6 (cylinder diameter in inches) × 3.1416 (value of pi) × 7 (teeth per inch) × 88 (rows of teeth) = 11,611. If ten seconds were required to insert one needle and bend it forward and another ten seconds were required to sharpen the needle with a file, then the time required for installing the 11,611 needles in the wooden cylindrical was 1,935 minutes, or about four days of continuous work.

In our tests, we found that we could reduce the sharpening time by flattening both ends of each needle before inserting it into the cylinder. That way, we only had to file the top side of the flattened tip, rather than filing the entire circumference, to create a point. Nevertheless, our analysis of Whitney's design also showed that the bent needles twisted sideways, came loose, and were thrown out by centrifugal force or pulled out by reluctant cotton seeds and became embedded in the unginned cotton. A gin built with square nails that had to be bent, shaped, and sharpened by hand[65] would have taken additional time to make, or about ten days per cylinder. In either case, it would probably take another ten man-days to fabricate the rest of the gin, or about one man-month per gin.

In 1823, John S. Skinner, editor of *The American Farmer*, described Whitney's gins in detail. According to Skinner, the cylinders were six to nine inches in diameter and two to five feet long. The surface of the rotating cylinder bristled with annular rows of bent-needle teeth (bent forward at an angle of about fifty-seven degrees), and the space between the teeth in each row was less than the width of a cotton seed. The rows of bent teeth were spaced more widely than the length of a cotton seed and matched the slots in the breastwork.[66]

A correspondent of Mr. John D. Legare, editor of the *Southern Agriculturalist*, wrote the following about studying Whitney's gins at Upton's Creek, Wilkes County, Georgia:

> The gins were water powered but still equipped with bent needles. After the bent-needle teeth were straightened and sharpened with a file, the gin operated tolerably well, but the impetus of the operation was too great for the substance they were attached to (wooden logs), which gave way, the teeth would fly out in the midst of the work and on occasion cause considerable trouble and loss of time.[67]

To increase production and reduce costs, Whitney may have considered modifying Oliver Evans' machine for manufacturing leather carding cloth by inserting long-legged staples into the back of a sheet of leather. The carding leather could then be wrapped around the rotating wooden cylinder with the staple teeth lined in an annular row, and the teeth could be bent forward by Evans' machine. Eighty-eight rows of bent, stable teeth could be quickly fabricated. Whitney may have learned mass manufacturing from Evans.[68] However, there is no evidence that Whitney used Evans' system for making carding paddles and drums. Instead, he built a factory, fabricated thirty gins, and shipped them south in time for the 1795 fall cotton harvest.[69]

Whitney's letters and Georgia folklore show that Whitney spent little time improving his cotton gin. Whitney's only objective was to fabricate a gin that demonstrably worked and was therefore patentable. He handed a model over to the U.S. Patent Office on March 14, 1794, and was granted a patent before the design was fully developed. Whitney's impractical design incorporated serious defects, including those found in this engineering analysis and described by Skinner and Legare's correspondent. Modern gins incorporate many of Holmes' design features and none of Whitney's.

Other Studies of Whitney's Cotton Gin

D. A. Tompkins' book *Cotton and Cotton Oil* and Charles Bennett's book *Saw and Toothed Cotton Gin Developments* show copies of Whitney's March 14, 1794, patent drawings and specifications and Holmes' May 12, 1796, patent specifications, both certified by Secretary of State James Madison. They also show Tompkins' 1900 drawings of Whitney's and Holmes' gins.[70] Unfortunately, Holmes never submitted a drawing of his gin with his patent application. Bennett stated that Tompkins made full-size models of both cotton gins according to the patent drawings in the early 1900s. These models were on display at Clemson University in Clemson, South Carolina. As of May 2006, the model gins were separately on display in South Carolina museums.

Holmes' and Whitney's original patent documents and patent models were destroyed when the U.S. Patent Office burned in 1836. After the fire, the patent office called for interested parties to restore their ancestors' patents by May 2, 1841. Whitney's descendants answered the call to promote Whitney's legacy. In addition to the original drawings, the descendants added drawings of two bent-needle cylinders with different-size wire teeth and three circular saw blades with different-size rip-saw teeth to the drawings of the restored patent. The draftsman also placed the handle that operated the gin on the doffer axle instead of the wooden cylinder axle. Holmes' patent was not restored.

Engineer Charles A. Bennett, in charge of cotton ginning investigations for the USDA Bureau of Agricultural Engineering in Stoneville, Mississippi, studied cotton gin design and operation throughout his career. On November 15, 1932, Bennett wrote to Mr. J. A. Humberstone of the Edison Institute in Dearborn, Michigan, concerning Whitney's 1823 model of a saw cotton gin, which is now displayed in the Henry Ford Museum and will eventually be discussed in this book. Bennett's letter summarized his feelings as follows:

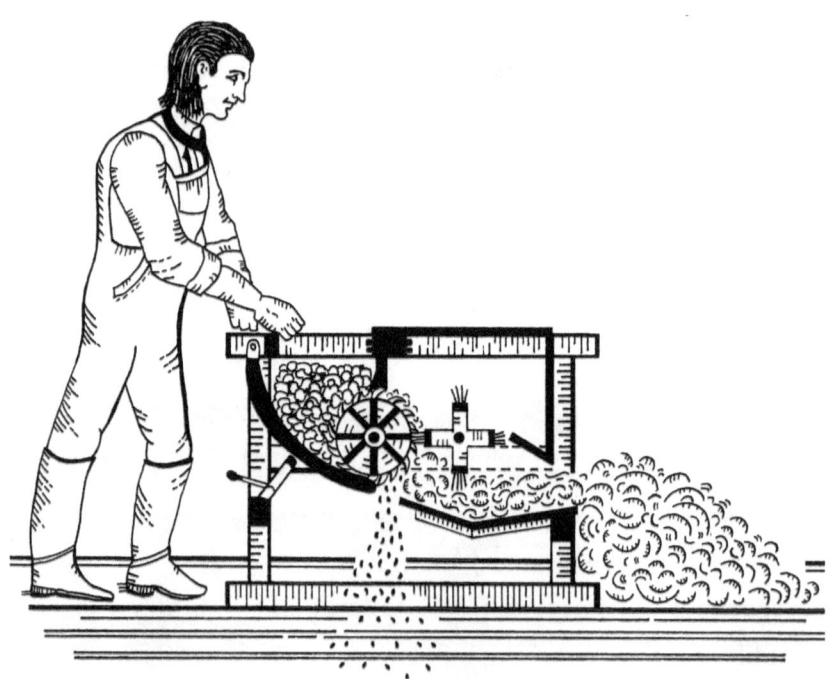

In about 1900, Daniel Tompkins, professor of mechanical engineering at Clemson University, reconstructed Whitney's bent-needle-toothed cotton gin. *Lorraine Kuehnel*

We do not wish to go on record as saying that the cotton ginning basic principle was not invented by Mr. Whitney, as we firmly believe that it was. Our observation being that the introduction of the saw teeth, which formed a very valuable improvement to the basic idea, was made by Holmes. In this matter, we, therefore, do not wish to detract anything from Mr. Whitney's deserving record and yet at this time [we] wish to give credit to Mr. Holmes for his valuable contribution.

Bennett could not accept that Whitney's and Holmes' gins were different in their design and operation and therefore merited two different patents. On the other hand, in his 1916 book, *Cotton as a World Power*, James Scherer was more forthright in his appraisal of the differences between the two gins. He wrote that the bent-needle teeth in Whitney's gin "tore open the seeds and badly mangled the fibers while oil, oozing from the crushed seeds, soon clogged the gin with moist fibers. Holmes eliminated the problem with saws which cut the lint free."[71] At the end of the 1808 patent litigation trial, the presiding judge declared that the two gins were, "in the eye of the law one and the same machine," although neither he nor Whitney understood that there were patentable differences between the two machines. In tacit confirmation of this, Whitney ultimately adopted the saw-tooth cylinder as an improvement and unashamedly assumed the credit for inventing it.

CHAPTER 4

The Patent History

Once Mrs. Greene showed others how to doff the lint off the rotating bent-needle cylinder of Eli Whitney's cotton gin, Phineas Miller urged him to quickly write a patent application for it. Miller knew that others were also fabricating cotton gins. Miller and Whitney wanted the honor of patenting the first cotton gin.[72] They also knew that patents would give the inventor a monopoly in the cotton gin industry.

After writing the patent application, Whitney mailed the specifications and drawings to Edmond Randolph, U.S. Secretary of State and member of the patent committee. He then began fabricating a model cotton gin, as required by the patent application process. After Whitney showed the finished model to his friend Joseph Stebbins, he hand-carried it to Philadelphia and presented it to the patent committee, consisting of President George Washington, Secretary of State Edmond Randolph, and Attorney General William Bradford, on March 14, 1794.

Several oddly interconnected activities took place in the U.S. Patent Office on that day. First, the caveat of invention for Hodgen Holmes' rip-saw-toothed cotton gin, which had been granted by the War Office on March 14, 1789, expired on that day, exactly five years after it was granted. The patent office should have upgraded Holmes' caveat of invention into a patent, with a life of nine more years, but chose instead to let it expire. Second, Washington certified on March 14, 1794, that Whitney's gin represented a new and different, and therefore patentable, idea. Finally, the U.S. Patent Committee backdated the granting date of Whitney's patent to November 6, 1793, the date on which the patent application had been received in Philadelphia.

General Nathanael and Catharine Greene were highly respected in the United States and good friends with many people in high places, such as George and Martha Washington, and Thomas Jefferson. So of course, they introduced Eli Whitney to their friends. Having friends in high places might have been why Whitney secured and protected his patent while Holmes' caveat of invention was allowed to expire. William Bradford, commissioner of patents, knew about Holmes' caveat of invention and chose to let it lapse. Also Thomas Jefferson exhibited extraordinary interest in the cotton gin when Whitney visited him to discuss the patent process on June 20, 1793. Similarly, George Washington was interested in acquiring a gin, as he hoped to diversify his tobacco plantation by growing cotton as well.

Miller and Whitney Begin Manufacturing Cotton Gins

In 1794, Phineas Miller and Eli Whitney, now partners in the newly formed M & W Company, organized themselves for business. They agreed that Whitney would manufacture his patented cotton gins in New Haven, Connecticut. They selected New Haven because the skilled labor and industrial supplies necessary for a manufacturing operation were more readily available there than in Savannah, Georgia. While Whitney supervised the manufacturing,

Miller would deal with southern planters in Savannah. They both knew their territories.

The M & W Company's partners planned to monopolize both cotton gin manufacturing and cotton ginning business. To do this, they would buy seed cotton and gin it themselves, and they would gin seed cotton brought in by planters, charging a fee equivalent to 40 percent of the cleaned cotton produced. For example, from three hundred pounds of raw seed cotton, brought in by a planter, the gin would produce about one hundred pounds of cleaned cotton. The planter would then take sixty pounds of cotton home, and the M & W Company would keep the remaining forty pounds as the ginning fee. The seeds extracted from the cotton were considered worthless at that time.

In addition to buying cotton outright and ginning it themselves, the M & W Company also planned to license gins to others, and to reinforce this strategy Miller and Whitney advertised that they would confiscate all unlicensed cotton gins. This would effectively prevent anyone else from ginning cotton. They also planned to prevent other companies from manufacturing cotton gins by filing patent infringement suits. In short, Miller and Whitney were going to exercise all of their legal training, and the patent rights granted by the federal government, to control the cotton market and establish a ginning monopoly.

Monopolies were odious to Americans, who had just fought a war in part to abolish British monopolies and the unfair trade acts. Unfortunately, for M & W Company, it lacked the manufacturing capacity to keep up with the demand for gins that was created by the ever-increasing cotton production in the United States. In addition, the company never had sufficient financial resources to buy an entire year's cotton crop outright. Nevertheless, M & W

Company's business plans did not endear it to its potential customers, the cotton planters. Georgia folklore says that cotton planters broke into the M & W Company offices several times to steal models of cotton gins.

In the winter of 1793–1794, Miller advertised that farmers should plant green-cotton seed in the spring and that the M & W Company would gin all delivered cotton.[73] At the same time, Whitney began manufacturing batch-type, bent-needle gins in the New Haven factory. By the summer of 1794, he had finished six gins and shipped them south to Miller for testing in time for the fall harvest. Two gins were placed on the Mulberry Grove plantation, and two were sent to two other destinations in Georgia and South Carolina. Miller claimed that the animal-powered ginnery at Mulberry Grove cleaned one hundred pounds of cotton per day.[74]

Meanwhile, Edward Lyon, Hodgen Holmes, and a host of South Carolinians had been manufacturing saw gins since 1784. Whitney called them "pirated gins," but these gins were fulfilling the ever-increasing demand for ginning capacity in the South.

In 1795, English spinners claimed that the cotton ginned with Whitney's bent-needle gins became tangled because of knots (called "neps") in the fibers. Whitney said that the cotton that contained neps was simply bad cotton. Sales of bent-needle gins fell in 1796, and Whitney had to carry ten unsold machines over the winter. They gathered dust in the warehouse.

In effect, the M & W Company faced quality control problems. Miller wrote to Whitney that merchants in England had examined samples from one hundred bags of cotton produced by Whitney's gins. The merchants found that irreparable damage had been done to the fibers. The damage had reduced the staple length and nearly destroyed the fibers' usefulness. Several spinners were even planning to buy cotton processed by roller gins, which produced cotton of more reliable quality, although much more slowly and in much smaller quantities.

In September 1797, Whitney complained to his father that the spinners in the city of Manchester were no longer buying cotton from his gins. These British spinners were used to spinning the fine, long-staple cotton from the Caribbean, and they had experience with brittle, short-staple cotton from India. However, when they imported American short-staple cotton that had been ginned by Whitney's needle-toothed gins, they complained that the gins had knotted the lint, making the cotton worthless for spinning. Whitney replied that the spinners had purchased poor quality, drought-stricken cotton. Similar comments are heard to this day in the modern cotton textile industry.

Within two years, the spinners had adjusted their equipment to accommodate different staple lengths and began buying short-staple cotton from Whitney's gins. The ginner's versus spinner's

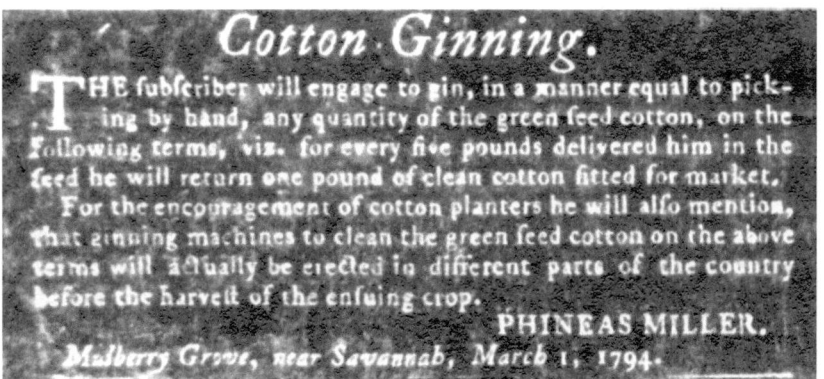

Miller ad

This ad offers Phineas Miller's ginning services to cotton growers near Savannah, Georgia. *Eli Whitney Papers (MS 554). Manuscripts and Archives, Yale University Library*

controversy had run its course, but permanent damage had been done to the M & W Company.

By the fall of 1797, the New Haven factory, which had been rebuilt after a 1795 fire, was nearly idle. Twenty-eight unsold gins sat on the shop floor collecting dust. The M & W Company was deeply in debt. Its monopolistic practices, exorbitant ginning fees, and court battles with other gin manufacturers and gin owners had alienated potential customers. Burdened with a flawed gin design and vilified for its businesses practices, the M & W Company collapsed into bankruptcy in 1798[75] after just three years of operation.

Whitney-Miller ledger

This accounting ledger shows the debts Eli Whitney took on once he ended his cotton gin–producing partnership with Phineas Miller.

Eli Whitney Papers (MS 554). Manuscripts and Archives, Yale University Library

In its final year, the M & W Company had only 30 gins operating at eight locations in Georgia. With the help of historical documents, we estimate that Whitney manufactured and shipped a total of 201 gins between 1794 and 1808.[76] At the same time, pirated gins (as Whitney called them), including Holmes' gins, were operating all over the South. These gins, not Whitney's, were ginning the bulk of the U.S. cotton crop.[77]

The failure of the M & W Company also bankrupted the company's patron, Mrs. Catharine Greene, the widow of General Nathanael Greene and, at the time of the bankruptcy, the wife and business partner of Phineas Miller. The Mulberry Grove plantation was sold at auction to pay debts and the Greene family moved to more modest accommodations to their other property on Cumberland Island, off the coast of Georgia.[78]

In early 1800, to meet their urgent need for cash, Miller and Whitney changed their tactics and pursued the sale of manufacturing licenses to cotton-growing states. This would allow others to manufacture cotton gins on strictly limited terms. The legislatures of three cotton-growing states purchased state manufacturing licenses from Miller and Whitney for a total of $90,000. South Carolina paid $50,000 on December 20, 1801; North Carolina paid $30,000 in 1802; and Tennessee paid $10,000 in 1802. These licenses gave manufacturers within these states the legal rights to manufacture cotton gins, which included both Hodgen Holmes' saw-toothed design and Eli Whitney's bent-needle-toothed design, royalty free.[79] Before the vote in the South Carolina state legislature, Holmes lobbied against purchasing the state license.

As a part of the license agreement, Whitney agreed to provide the state of South Carolina with two prototype cotton gins.

However, being busy with his new business of manufacturing muskets, Whitney did not immediately fabricate these gins. After two years had passed without delivery of the gins, South Carolina canceled the licensing contract and sued Whitney to recover the down payment. In due course, Whitney designed and manufactured two full-size cotton gins. Unfortunately, the design, specifically the tooth and hopper configuration, of these gins is not known.

Miller and Whitney Defend Their Patent

As soon as Holmes learned about Whitney's far-reaching 1794 patent, and Whitney's aggressive litigation against other cotton gin manufacturers and ginners, he wrote a patent application for his own earlier saw cotton gin and filed the application on April 19, 1796. George Washington certified on May 12, 1796, that Hodgen Holmes had invented the first cylindrical, continuous-flow, rip-saw-toothed cotton gin equipped with spaced metal saw blades and near-vertical ribs. Washington certified that this was a new, patentable idea, and Holmes was granted a patent with a life of fourteen years.

We believe Washington was qualified to evaluate this invention. He saw that Holmes' saw-toothed gin with vertical ribs was indeed a new and different invention compared to the bent-needle-toothed cotton gin with a tangential wooden breastwork. However, Washington may not have also understood that Holmes' gin was an efficient, continuous-flow design, while Whitney's was a less practical, batch-type gin.

Meanwhile, Miller and Whitney continued to aggressively defend their patent. Their policy seemed to be: File enough lawsuits and you will eventually find a judge who will agree with you. However, their long record of litigation also teaches that the royalties from a patent are often far less than the cost of protecting the patent. Miller and Whitney used the $90,000 they received for the sale of state manufacturing licenses to bankroll their ongoing patent protection efforts in the 1800s. As Whitney wrote to his father, "We get but a song in comparison to what it's worth." Their efforts also included maneuvering to change the rapidly evolving patent laws. Fortunately for posterity, one patent rule that they promoted has not been adopted: "All improvements to an invention belong to the senior filer of the original invention." They also promoted another patent rule that worked in their favor and that still exists: "The first to file a patent application is senior to any later filing by other inventors."

Starting in 1795, after the Whitney patent was granted, until his death in 1804, Phineas Miller filed as many as seventy-two lawsuits against perceived patent infringement, typically seeking triple damages. For example, an Edward Lyon of Wrightwood, Georgia (twenty miles south of Augusta), built a water-powered, needle-toothed, batch-type gin and sold it to William Kennedy. Lyon had not applied for a patent or for a license from the M & W Company. Thus, Phineas Miller's first two suits were filed in 1795 against Lyon and against Kennedy.[80]

In a letter dated February 15, 1797, Miller informed Whitney that a Robert Holmes (the authors believe Whitney was mistaken about Holmes' correct first name) was manufacturing and selling "ratchet wheel toothed" cotton gins. In this letter, Miller and Whitney showed their true regard for the legal process. Miller requested that Whitney depose his friend Joseph Stebbins and another witness named Goodrich on the subject of the ratchet wheel design. Miller asked Whitney to instruct Stebbins and Goodrich that each should testify by written deposition that

"Whitney repetitively told me that he contemplated making a whole row of teeth from one plate of sheet metal." Stebbins, then a judge in Alms, Maine, was known for his integrity, and Goodrich did not even respond to Whitney's proposal. They did not want to perjure themselves.[81]

In later cases, Whitney testified several times that he had considered using circular saw blades. This is an admission, we believe, that Whitney had studied saw-toothed cotton gins during his travels through South Carolina, including a visit to the Kincaid plantation in the winter of 1792–1793, where he saw Holmes' saw gin, as claimed by South Carolina folklore.[82] Nevertheless, Whitney wanted the court to believe that the rip-saw-toothed gin had first appeared when Hodgen Holmes filed his 1796 patent application, even though South Carolina began shipping ginned cotton lint as early as 1784. By appropriating Holmes' rip-saw-tooth design as merely an improvement of his own patented gin, Whitney became Holmes' nemesis and vice versa.

In effect, Miller and Whitney were legal opportunists. They manipulated the legal system, the U.S. Patent Office, and the U.S. District Circuit in Savannah, Georgia. In the fall of 1793, they requested that the secretary of state disregard any patent applications submitted for improvement to their original design. They continually pressed for the establishment of patent rules that would benefit their situation. In 1801, they successfully lobbied the U.S. Congress to change the patent laws.

Three years later, Miller and Whitney filed their twenty-fourth lawsuit, this one against Arthur Fort and John Powell in U.S. District Circuit in Savannah. This case involved the question of whether Holmes' gin represented an infringement on Whitney's patent. During the preparation for the trial, Whitney hired a draftsman to duplicate the drawings and specifications of his 1793 patent and of Holmes' 1796 patent, both of which were filed in the U.S. Patent Office. He requested Secretary of State James Madison to compare the original documents with the draftsman's reproductions. Madison certified the copies as true copies. The drawings were filed on April 27, 1804, with the Savannah court and were subsequently used in cases tried in 1804, 1806, and 1807. Fortunately, these handwritten, certified copies still exist among the litigation records in the U.S. District Courthouse in Savannah. The drawings and specifications of Whitney's gin pertain exclusively to a bent-needle cotton gin.[83]

In the Whitney vs. Carter case of 1806, defense witnesses testified about the earlier use of circular saw blades, protected by slotted ribs, for removing the husks from coconuts in the Caribbean and for deburring wool in Ireland,[84] and witnesses testified that Holmes may have reinvented the circular saw blade and slotted rib design.[85] Nevertheless, Whitney argued in court that the claims of his patent for the bent-needle gin also covered the claims of Holmes' patent for the saw-tooth gin.

Whitney next sued Isaiah Carter and Charles Gachet for using a saw gin. The accused denied the charge. They claimed that since Whitney had not invented the saw gin, they could not infringe on a nonexistent patent. The defense rested its case on this argument. The court found for the plaintiff, and damages were awarded against Isaiah Carter for $2,000 and against Charles Gachet for $1,500.[86]

On December 19, 1806, the court rendered a verdict for the plaintiff in the twenty-fourth lawsuit, Whitney vs. Arthur Fort and John Powell. The U.S. District Court found that Whitney's patent rights for a bent-needle gin were senior to Holmes' saw gin.

In finding for Whitney, Judge William Johnson called Whitney's patent "the basic cotton gin patent." He further declared that Holmes' saw-toothed gin was merely an adaptation of the bent-needle-gin and was therefore covered by the claims in the Whitney patent.[87] He also said that bent-needle teeth were the same as saw blade teeth. The judge never addressed the differences between a batch-type gin and a continuous-flow gin.

But the judge went too far. The Savannah Federal Court was becoming weary of Miller and Whitney's endless litigation against cotton gin manufacturers and cotton ginners. The court began asking what Whitney really wanted, and it appears that the court was ready to satisfy him. In finding for the plaintiff in an earlier case in 1802, Judge William Stephens had issued a nullification of Holmes' patent on the grounds that the bent-needle-toothed gin and the rip-saw-toothed gin were "in the eyes of the law, one and the same gin." In 1807, Judge William Johnson upheld Stephens' 1802 decision. He again nullified Holmes' patent and added an injunction against Holmes' estate, as requested by Whitney. The injunction prevented Holmes or his assignees from manufacturing and selling rip-saw-toothed cotton gins. Ironically, Stephens' and Johnson's rulings were of little monetary value to Miller and Whitney. Whitney's original patent expired on November 6, 1807.

Whitney Secures His Legacy as Inventor of the Cotton Gin

In 1800, Whitney made five models of cotton gins for use in court cases, in an attempt to secure his legacy as "the" inventor and to establish his ownership of the saw gin design.

Whitney never understood the principles of Holmes' cotton gin, yet through the courts he was able to secure control of the saw-toothed cotton gin patent. Whitney used his appearances in court to reinforce his claim. For example, during the twenty-fourth patent infringement case, Whitney displayed a model cotton gin that incorporated two types of teeth: six rows of saw teeth and twelve rows of bent-needle teeth. Whitney demonstrated the model in court by ginning a handful of seed cotton. Whitney first fed seed cotton into the bent-needle section and cleaned cotton emerged from the gin. Then Whitney fed cotton into the saw-toothed section and, again, cleaned cotton emerged from the gin. This was a clever way of showing that bent-needle teeth were interchangeable with circular saw blades. Whitney was careful not to operate the gin long enough to fill the hopper with ginned cotton seeds, which were stripped from the cotton and returned to the hopper because of the nature of Whitney's gin design. This would have shown that the gin was a batch-type design. Consequently, the judges did not understand how vertical ribs and a specially designed hopper promoted a continuous-flow ginning process.

Whitney died on January 8, 1825. Even toward the end of his life, he felt the need to promote his legacy as inventor of the cotton gin. While still active, in 1823, in another attempt to secure his legacy as inventor of the rip-saw-toothed cotton gin, Whitney constructed a model of a batch-type gin with a tangential breastwork and horizontal apron. He included saw teeth in this model.[88] Whitney fabricated a model gin with twelve circular saw blades. This model is currently part of the collection of the Henry Ford Museum in Dearborn, Michigan.

A study of this 1823 model shows a wooden breastwork and horizontal apron are tangent to and lay on top of the cylinder rather than being vertical at the side of the cylinder, as in Holmes' design. This design could not develop a rotating seed roll and dispose of the stripped green seeds. Thus, this was an impractical, batch-type gin. Whitney also made other models of saw gins to be exhibited in museums, including a model given to the Smithsonian National Museum, in Washington, D.C., by Eli Whitney Jr. in 1884.[89]

In 1959, while employees of the New Haven Colony Historical Society were emptying a storage room in the basement of their headquarters building, they discovered a long-forgotten box. On closer examination, they found the box contained a disassembled Whitney rip-saw-toothed cotton gin. A tarnished metal plate on the side of the gin was inscribed, in large letters: "Whitney's Patent." On the other side were the words: "Invented 1793, Made in 1803." Painted underneath the hopper was the year, "1807."[90] This gin and several other gins are now displayed by the New Haven Colony Society Museum in Connecticut.

The cylinder of this 1807 gin consisted of forty saw blades. Each blade was 6.75 inches in diameter and had 106 teeth. The blades were separated with 0.75-inch tin or pewter spool-shaped castings. A line drawing made by the New Haven Colony Historical Society shows that the horizontal breastwork is tangent to the saw blades and that the hopper would therefore not promote the creation of a seed roll. This was, therefore, a batch-type gin, like all of Whitney's gins. However, because it contains saw blades rather than bent-needle teeth, it does not conform to the specifications of Whitney's original patent and instead reflects Whitney's attempted assimilation of Holmes' superior design.

The New Haven Colony Historical Society concluded that the gin was constructed around the time that Whitney was engaged in fulfilling an order for seventy-five to eighty gins from the State of South Carolina. We surmise that South Carolina placed this order to provide local craftsmen with reference models for duplicating Whitney's cotton gins for sale to planters. This could have been done under the terms of the state's manufacturing license. Whitney may have made an extra gin and stored it in his factory.

The inventory of Whitney's estate in 1825 listed "stores" for muskets as well as "99 boxes" containing cotton gins. In addition, the catalog for the 1854 Industrial Fair in New York mentions that a Whitney cotton gin was on display. The New Haven Colony Historical Society apparently acquired this gin after the exhibition and stored it, along with various other historical artifacts, in the basement of its headquarters building,[91] where it was eventually forgotten. It is now on permanent display.

The design of Whitney's 1807 gin is similar to Tompkins' cross-sectional drawing of 1901.[92] They were both batch-type gins, and they both contained horizontal breastworks. The size and shape of the hopper were not conducive to developing a rotating seed roll, which was one of the features of Holmes' design and is still used in all modern gins.

Whitney Turns to the Manufacture of Muskets

When Phineas Miller died on December 7, 1803, he left Whitney to manage the ongoing court battles in Savannah and Whitney's new enterprise of manufacturing muskets for the U.S. Army in New Haven, Connecticut. Despite their aggressive litigation, the ginning business was being taken over by "pirated gins," and Whitney had been looking for other manufacturing opportunities. He answered a newspaper advertisement placed by President John Adams' administration calling for the submission of a contract for manufacturing muskets for the U.S. military. In 1798, Whitney submitted a contract for manufacturing ten thousand stands of arms (a "stand of arms" consisted of a musket, bayonet, ramrod, wiper, and screwdriver).

He specified in the contract that he "would mass manufacture stands of arms that would have the quality of interchangeable parts" and he would deliver ten thousand stands of arms in just two years. Thomas Jefferson supported the two-year contract. Whitney manufactured ten thousand stands of arms, equipped with interchangeable parts, and delivered the final shipment in January 1809, ten years late, but just in time for use in the War of 1812. During that long interval, Whitney is credited with inventing the American way of manufacturing by using mass-produced, interchangeable parts. After the war, Whitney received another contract for fifteen thousand more stands of arms, and he became wealthy on government contracts.

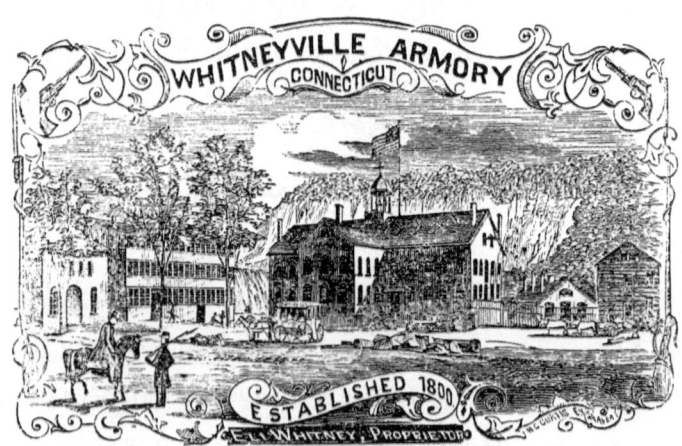

WHITNEY'S IMPROVED FIRE-ARMS.

Whitney musket factory ad

Whitney Arms Company advertisement, 1860. Eli Whitney began manufacturing muskets for the U.S. Army before his cotton gin operations completely ended. *Library of Congress*

Aside from Whitney's dubious claims for invention of the cotton gin, we should be grateful for Whitney's achievements in manufacturing.[93] However, in 1966, doubts arose about Whitney's claimed "interchangeability of parts" for his mass-produced muskets. In theory, a number of muskets could be disassembled, their parts mixed, and the same number of identical muskets could then be reassembled from the randomized parts. However, Edwin A. Battison of the Smithsonian Institution scrutinized muskets made by Whitney that are now on display in the New Haven Gun Museum. He found a Roman numeral VI engraved inside the lock of a Whitney musket. Following this clue, Battison determined that the musket parts were not uniform; they had been individually worked by hand and were therefore not truly interchangeable.[94]

CHAPTER 5

The Evidence in Cotton Production

It's interesting that Denison Olmsted, who promoted Eli Whitney's legacy as inventor of the saw gin, inadvertently destroyed that legacy by publishing data on cotton shipments from Charleston, South Carolina, to Liverpool, England, from 1784 through 1788. The data indicate that rip-saw-toothed cotton gins were operating in the South and that ginned cotton lint was being shipped to Liverpool from Charleston and other ports before Whitney invented and patented his cotton gin.[95]

The data support the evidence that we have accumulated and support our belief that Whitney was not the original inventor of the saw gin. He was only the inventor of an impractical bent-needle-toothed, batch-type cotton gin. He manufactured his design of a cotton gin for three years and his company, the M & W Company, became bankrupt. He also bankrupted his patron, Mrs. Catharine Greene.

The following table shows increasing shipment of cotton from Charleston, South Carolina, and other ports, from 1784 to 1880.

Table 5-1
Production, Export, and Sale Price of
U.S. Green-Seeded Cotton Lint, 1784–1880

| Year | Cotton Production | | Export | Price |
	bales	lbs		
1784	--	2,000	8 bags	--
1785	--	1,250	5 bags	--
1786	--	1,500	6 bags	--
1787	--	27,000	108 bags	--
1788	--	70,500	282 bags	--
1789	--	--	--	--
1790	3,130	1,569,000	--	--
1791	4,154	2,092,000	--	--
1792	6,276	3,138,000	--	--
1793	10,460	5,230,000	--	--
1794	16,736	8,368,000	--	35¢/lb
1795	16,736	8,368,000	--	35¢/lb
1796	20,921	10,760,500	--	37.5¢/lb
1797	23,013	11,865,500	--	40¢/lb
1798	31,381	15,690,500	--	
1799	41,841	20,920,000	--	
1800	73,222	35,611,000	--	44¢/lb
1801	100,418	50,209,000	--	--
1802	115,063	57,531,000	--	--
1803	125,523	62,764,500	--	--
1804	135,983	67,991,500	--	--
1805	146,444	73,222,000	--	31¢/lb
1806	167,364	83,682,000	--	--
1807	167,364	83,682.000	--	--

1808	156,904	78,452,000	--	--
1809	171,548	85,774,000	--	--
1810	177,824	89,912,000	--	--
1811	167,364	83,866,000	--	--
1812	156,904	78,452,000	--	
1815	209,205	104,602,000	--	--
1820	334,728	167,364,000	--	--
1825	533,473	266,736,500	--	--
1830	732,218	366,109,000	--	--
1835	1,061,821	530,910,500	--	--
1840	1,347,640	673,200,000	--	7¢/lb
1845	1,806,110	903,055,000	--	--
1850	2,136,083	1,068,041,500	--	--
1860	3,841,416	1,923,208,000	--	--
1870	4,024,527	2,012,263,500	--	--
1880	6,356,998	3,178,499,000	--	--

Note: Shipping data for 1784–1788 are from Denison Olmsted's *Memoir of Eli Whitney, Esq.* (1832). The rest of the shipping data are from James Augustin Brown Scherer's *Cotton as a World Power* (1916). Note that there was little increase in cotton shipments during the first few years Whitney was manufacturing or exercising his patent rights concerning his cotton gins. He exercised his rights to prevent others from using, manufacturing, or selling his gin.

Southern growers of rice, indigo, and tobacco became excited about the prospect of growing and ginning the hardier and more prolific American green-seeded cotton when they heard about the prosperity of plantations that used the new saw-toothed cotton gins. For his

1796 cotton crop, James Kincaid ginned 33,749 pounds of cotton and received 37.5 cents per pound, for total of £12,656. In total, 9,500,000 pounds of ginned cotton were shipped by the United States that year. Therefore, Kincaid personally shipped 0.36 percent of the total ginned cotton shipped from U.S. ports in 1796. The Kincaid plantation had high ginning capacity and was a profitable plantation.

The accounting records of George Foster, bookkeeper, show that the Kincaid plantation was growing cotton, buying seed cotton from neighboring planters, and ginning cotton for shipment to Charleston in 1794. That was the year Whitney shipped six experimental gins south from his new factory in New Haven, Connecticut, for testing in South Carolina and Georgia. There is no evidence that Whitney gins were shipped to the Kincaid plantation. The Abstract of Deeds for Fairfield County shows that James Kincaid began purchasing land nearly every year after 1787, and he possibly began even earlier than that. We believe that the money for these acquisitions came from the sale of cotton lint.

In 1793, Phineas Miller urged Eli Whitney to hurry the patent application for Whitney's cotton gin because he knew there were other inventors who could also patent their own cotton gins. Whitney constructed a gin as fast as possible, but he could not find steel plate with which to make saw blades, as he had seen in South Carolina. Therefore, he submitted a wooden log, bent-needle-toothed gin design to the U.S. Patent Office on March 14, 1794, and was granted a patent. After this patent was granted, the M & W Company initiated lawsuits against gin inventors, manufacturers, and operators and any other potential infringer of its patent. Miller cautioned Whitney that others blacksmiths were building saw gins and might not know that Whitney had been granted a patent that covered their gins.

By 1795, there was no increase in cotton shipment from the United States. In that year, only about thirty-six Whitney gins were operating in the South. These gins would have only ginned about 1,440 bags in the fall of 1795, while local blacksmith-fabricated saw gins processed 30,550 bags of cotton. Cotton was being shipped in woolen marketing bags used for shipping wool. The bags held approximately 250 pounds of ginned cotton.

Over the next two years, shipments of cotton from the United States only slightly increased. Then from 1798 and 1799, shipments of ginned cotton went up drastically, after Whitney had begun manufacturing muskets. While these cotton crops could have been finger-ginned or churkha-ginned, it is more reasonable to believe that rudimentary saw gins were in use, having been invented by local craftsmen or blacksmiths such as Hodgen Holmes. We believe the invention of the cotton gin may have been promoted during the Revolutionary War to provide clothing for the troops. Hodgen Holmes was the only southern blacksmith who filed a patent application for a caveat of invention for a cotton gin to the U.S. War Office.

Assuming that the first saw gins were fabricated after the circular saw blade was invented in England or reinvented in the United States, we believe that rip-saw-toothed gins were being fabricated in South Carolina and/or Georgia before 1784 because cotton was being continuously shipped to Liverpool from Charleston and other seaports[96] just after the Revolutionary War ended in 1783. We also believe that this was green-seeded cotton because black-seeded cotton was introduced to Georgia several years later, in 1786. This quantity of cotton could not have been finger-ginned or churkha-ginned. It had to have been ginned with mechanized saw gins.

The M & W Company held the cotton industry hostage while defending its overreaching patent. Nevertheless, cotton production increased enormously during the formative years 1784 to 1795 and after 1815. Historians have observed that annual shipments of green-seeded cotton from the South nearly doubled every year during 1784 to 1795. Table 5-1, on pages 96–97, shows that cotton was being grown and ginned in the United States and shipped starting in 1784. Unfortunately, this production information provides little information on the earliest development of the cotton industry. In addition, South Carolina folklore is silent about the earliest exports of green-seeded cotton from South Carolina to Liverpool.

In his *History of Charleston County, South Carolina*, Thomas Petigru Lesesne wrote: "In the year of 1784, John Teasdale, a merchant of Charleston, shipped from the city to J. and J. Teasdale and Co., Liverpool England, eight bags of cotton." Olmsted then took up the story and wrote about this important historical event in his *Memoir of Eli Whitney* in 1832: "In 1784, an American vessel arrived at Liverpool, having on board, for part of her cargo, eight bags of cotton, which were seized by the officers of the customs house, under the conviction that this quantity of cotton could not have been grown in America."[97] Upon satisfactory proof of the cotton's origin, the cargo was released. Olmsted also listed the number of bags of ginned cotton shipped from Charleston, Philadelphia, and New York to Liverpool during the next four years: 5 bags in 1785, 6 bags in 1786, 108 bags in 1787, and 282 bags in 1788.

The quantity of cotton ginned and shipped from the South increased dramatically once village and plantation blacksmiths began fabricating rudimentary saw gins. In fact, the total weight of cotton shipped almost doubled each year during these formative years.

Another factor that probably led to the increases in cotton production was that the land had already been cleared for other crops, such as rice, tobacco, indigo, wheat, and corn, and was therefore ready for growing cotton. The planters owned the necessary equipment and manpower and they were looking for a new cash crop, in part because the British had shifted production of indigo to the East Indies during the Revolutionary War. In fact, after the war, the price of indigo was so low that southern planters readily shifted their indigo land into cotton production.[98]

In those formative cotton-producing years, production and shipment of cotton had some irregularities. In fact, the cotton industry never hit its stride until after Whitney's patent expired and he turned his attention to manufacturing muskets in 1812.

We analyzed the data presented in the table on pages 96–97 and concluded that the cotton production timeline could be divided into six segments (major periods). Our mathematical analysis and reasoning for the six segments are presented in Appendix 4. Here is a breakdown of those periods:

The First Segment: 1784–1790

Cotton was shipped under the following conditions:

1. There was an unlimited market for the cotton (in England and New England).
2. Land that had previously grown other crops was now available for growing cotton.
3. A supply of green-seeded cotton seed from local gardens was available for the first crop on plantations.
4. The planters were ready to grow cotton.
5. Local blacksmiths were ready to build saw gins and other necessary machinery.

6. There were no limiting factors such as money, labor, patent royalties, etc.

7. This was a period of rapid change and transfers of technology.

The continuous-flow, rip-saw-toothed cotton gin was probably invented before 1784, fabricated by local blacksmiths, and used for ginning green-seeded cotton. Hodgen Holmes was one of the blacksmiths who developed and manufactured the saw gin. He was also the only inventor who patented a saw gin. Although Holmes was granted a caveat of invention for his invention on March 14, 1789, he did not sue other manufacturers of saw gins. He made no

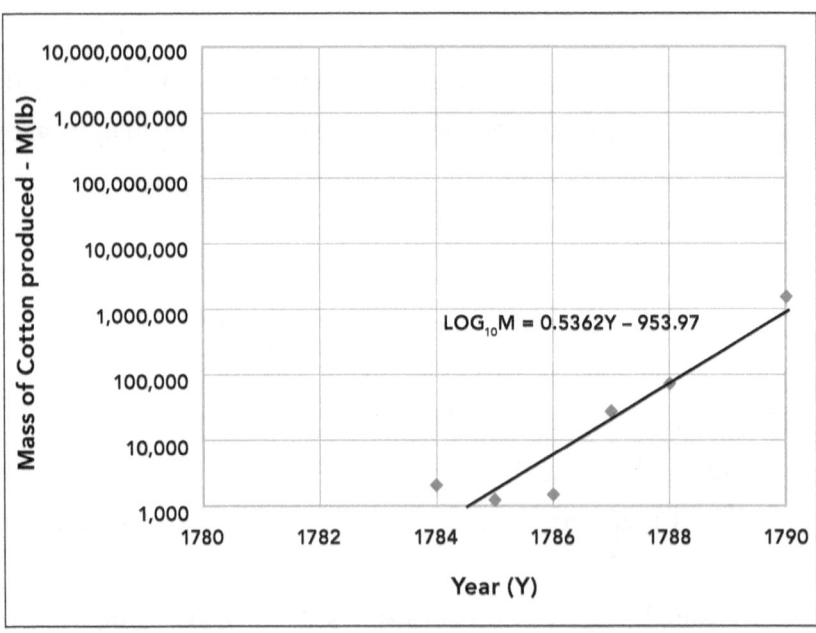

Graph 5-1. Annual shipment of cotton lint from the southern United States, 1784 to 1790 (plotted on semi-log scale). The increase in quantity of cotton lint shipped materially increased each year when compared to the total quantity shipped the previous years. It was a very healthy industry.

demands for patent royalties from other gin manufacturers, nor for ginning fees from ginners of cotton.

The Second Segment: 1791–1795

The U.S. Patent Office was established in 1790 and then revised in 1793. Eli Whitney graduated from Yale and traveled south in the fall of 1792. He studied saw gins and the ginning of cotton in South Carolina and Georgia. He fabricated a wooden-log, batch-type, bent-needle-toothed cotton gin because he did not have a supply of steel for making saw blades. He received a patent for his bent-needle cotton gin on March 14, 1794. This patent was back-dated to November 6, 1793, by the U.S. Patent Office.

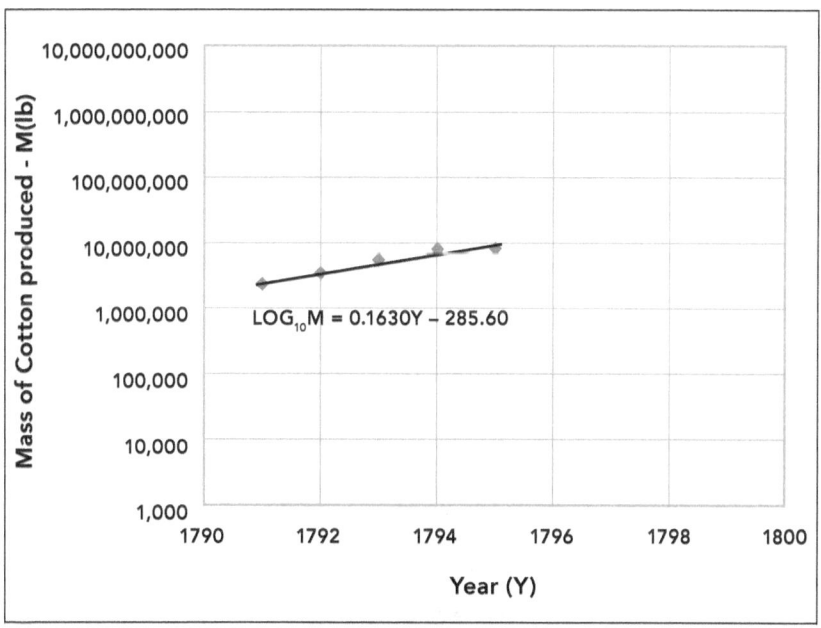

$$LOG_{10}M = 0.1630Y - 285.60$$

Graph 5-2. Annual shipment of cotton lint from the southern United States, 1791 to 1795 (plotted on semi-log scale). The slope of $log_{10}M$ materially decreased after the establishment of the U.S. Patent Office, but it was still a health industry in 1790 and became re-established in 1793.

Phineas Miller advertised in newspapers, in the spring of 1794, that the U.S. Parent Office had granted the M & W Company a patent for a cotton gin. He advised planters to plant green-seeded cotton in the spring and bring their cotton to the M & W Company's ginning stations that fall for ginning. He also advised owners of cotton gins to license their cotton gins with the M & W Company or their cotton gins would be confiscated and destroyed.

Georgia folklore states that Mrs. Catharine Greene showed Whitney's gins to various groups of people, and that someone immediately stole his model gin and began fabricating cotton gins. However, we believe that South Carolina blacksmiths began fabricating saw gins nine years before Whitney came south and fabricated his bent-needle gin in 1793. The increase of cotton ginning decreased during this second period.

The Third Segment: 1796–1802

Whitney began manufacturing cotton gins in New Haven, Connecticut, in the winter of 1794 and 1795 with the launch of the M & W Company. The building and twenty gins burned to the ground in March 1795. Whitney rebuilt and shipped thirty gins south in time for the fall harvest. During this period, the company began actively suing other gin manufacturers. Due to the publicity surrounding the patent trials, the annual increase in cotton ginning decreased from 1796 to 1802.

The Fourth Segment: 1802–1806

On November 6, 1802, Judge William Stephens issued the first patent judgment in favor of the M & W Company and nullified Holmes' patent. Holmes and other gin manufacturers were forced to purchase licenses from the M & W Company, starting in the

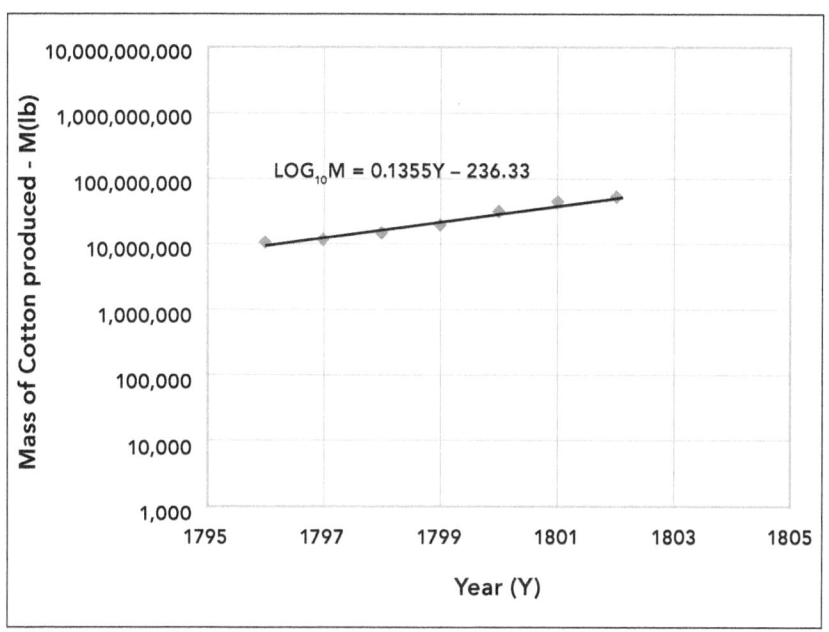

Graph 5-3. Annual shipment of cotton lint from the southern United States, 1796 to 1802 (plotted on semi-log scale). Again, while the total quantity of cotton shipped increased, the value of $\log_{10}M$ and the slope of the line decreased when compared with the preceding years due to lawsuits among cotton gin manufacturers and cotton ginners.
The industry was starting to suffer.

spring of 1803. The M & W Company began receiving more plaintiff verdicts and collecting triple royalty fines.

Whitney shipped bent-needle gins south, but Miller removed the wooden cylinder bristling with bent needles and installed a circular saw blade cylinder. Miller did not modify the hopper and the breastwork to include vertical ribs, which would have promoted continuous-flow operation and self-disposal of the seeds. Consequently, the modified gin was still a batch-type gin that had to be periodically stopped and emptied. The annual increase in cotton ginning materially decreased as other manufacturers nearly stopped manufacturing gins.

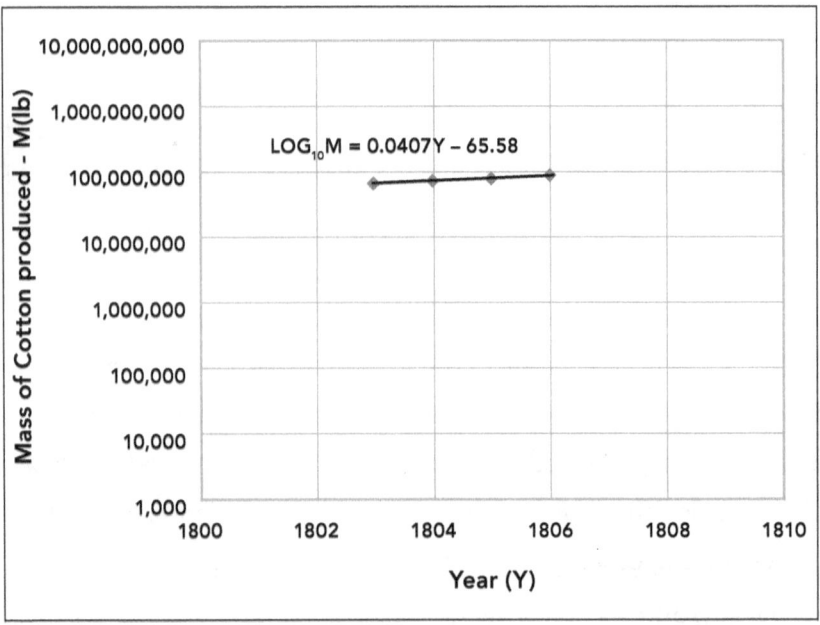

Graph 5-4. Annual shipment of cotton lint from the southern United States, 1803 to 1806 (plotted on semi-log scale). The total quantity of cotton shipped continued to increase. The proportional increase (*P*) decreased due to lawsuits filed by the M & W Company, and the value of $\log_{10}M$ materially decreased. The health of the cotton industry was deteriorating.

The Fifth Segment: 1807–1812

Thomas Jefferson enacted the ineffective Embargo Act of 1807 in December 1807, which forbid any American ship from leaving an American port for a foreign port. Exports bottomed out the following year. By the next year, U.S. officials realized the Embargo Act was a disaster and repealed it. Still, cotton exports were essentially flat until 1814. Then in 1815, exports increased and began a long expansion up to the Civil War.

Meanwhile, Whitney became busy with his War Office contract for fifteen thousand muskets. The increase in the growth of gin manufacturing became stagnant leading up to the War of 1812. In

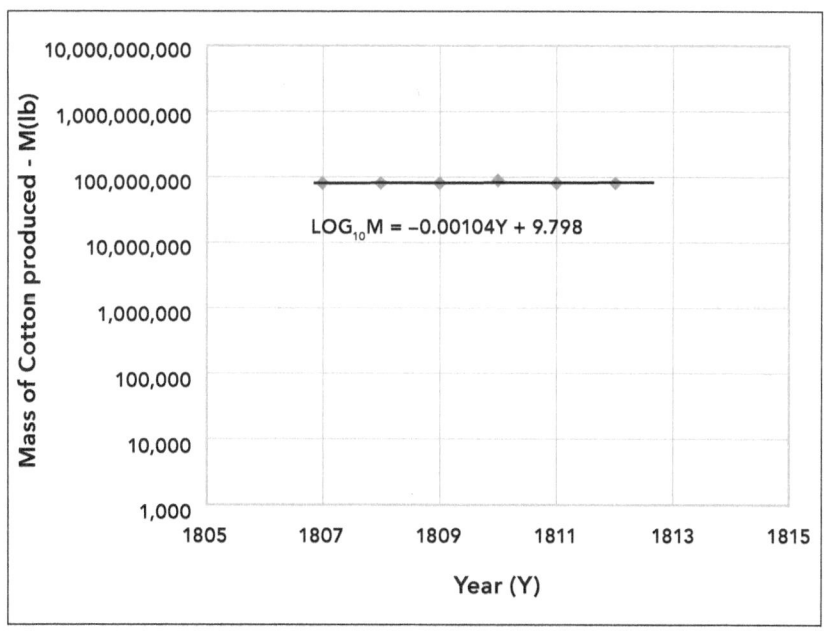

Graph 5-5. Annual shipment of cotton lint from the southern United States, 1807 to 1812 (plotted on semi-log scale). While the total quantity of cotton shipped slightly increased, the value of $\log_{10} M$ became nearly zero due to legal actions taken by the M & W Company and President Thomas Jefferson's embargo.

WHO REALLY INVENTED THE COTTON GIN?

addition, Judge William Johnson announced that the saw blade of the rip-saw-toothed gin was "merely a more convenient mode of making the bent-needle gin" and that the Whitney patent covered the saw blade.[99]

Judge William Stephens wrote in 1810 that Whitney's gin and Holmes' gin were in the "eyes of the law, one and the same machine."[100] Because the M & W Company was winning its suits and collecting triple royalty fines, it spread fear in the cotton gin industry after 1806. Whitney nearly shut down the growth of saw gin manufacturing in 1807. The proportional increase in growth dropped almost to zero. Whitney slowed the manufacturing of cotton gins starting in 1806; consequently, exports of cotton were low in 1807 and from 1811 to 1814.[101]

The Sixth Segment: 1812–1880

A study of the shipment of cotton lint shows that M & W Company's harassment of the saw-blade gin manufacturing industry lasted nearly twenty years. Then, Whitney became busy manufacturing an order for fifteen thousand muskets. After 1812, the increase of cotton production was nearly always positive for the next hundred years.[102] Luckily for the cotton industry, the U.S. Congress voted not to extend Whitney's 1793 patent in 1807 and again in 1811. Also luckily for his family, the U.S. government gave Whitney a contract for manufacturing fifteen thousand stands of arms after the War of 1812. Whitney died in 1825, wealthy from his government contracts.

During his last years, Whitney had no economic reason for continuing to harass cotton gin manufacturers or cotton ginners. Nevertheless, he worked on extending his legacy as the inventor of the saw gin. He fabricated a rip-saw-toothed gin that is now on display in the Henry Ford Museum in Detroit. It is a batch-type gin. It seems that Whitney must not have known how to make or manufacture a continuous gin!

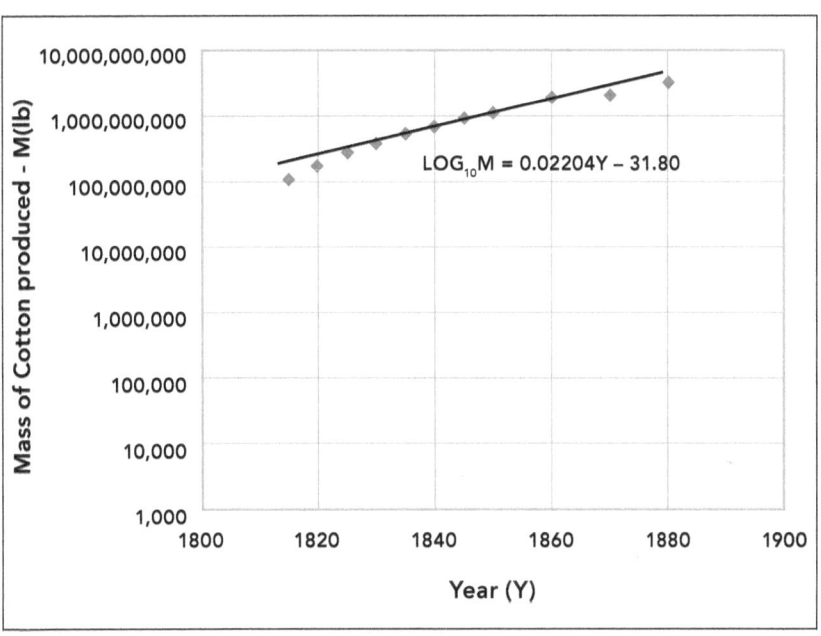

Graph 5-6. Annual shipment of cotton lint from the southern United States, 1812 to 1880 (plotted on semi-log scale). After Whitney's patent expired and after he received a contract for manufacturing fifteen thousand stands of arms, Whitney bowed out the cotton gin battle. As a result, the value of $\log_{10} M$ became positive for the next hundred years. The cotton industry was healthy again.

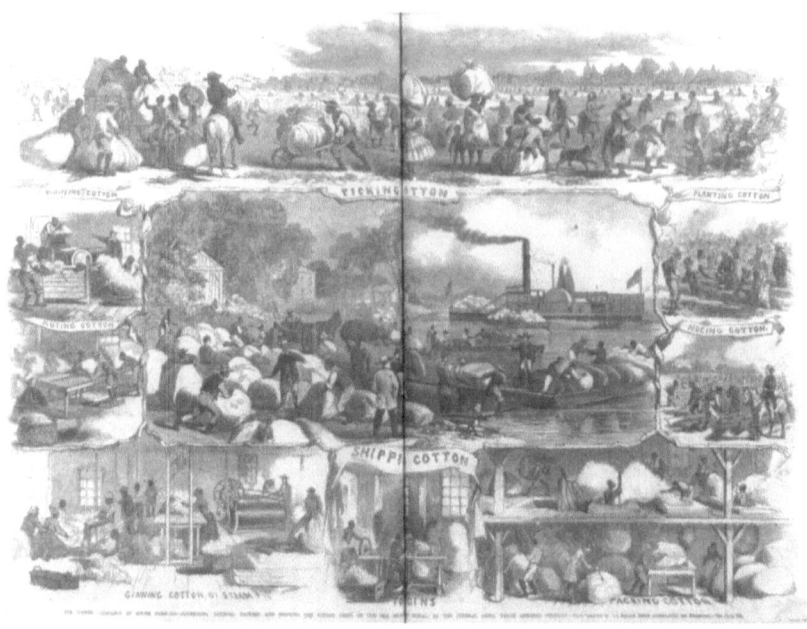

South Carolina cotton production advertisement

This composite of scenes comes from the Our Cotton advertising campaign of cotton producers in South Carolina. It features the packing and shipping crops from Sea Island in 1862. *Library of Congress*

Cotton picking

This cotton picking scene is from 1902 in Aiken, South Carolina. *Library of Congress*

Our Findings for the Years 1784–1880

Graph 5-7 shows that the proportional increase in growth of cotton production materially decreased after Whitney began collecting patent royalty triple fines from the manufacturers or ginners using saw gins. At the start, as shown in Graph 5-7, the increase in production of cotton was nearly double that of the previous years, but annual increase in production settled at 40 percent annual increases by 1802.

An increase in clean cotton production would depend on the increase in the capacity of and the number of saw gins. Therefore, any stagnation in the yearly increase in clean cotton production over time indicates that there were problems with the production of seed cotton or with the manufacture of saw gins. We postulate

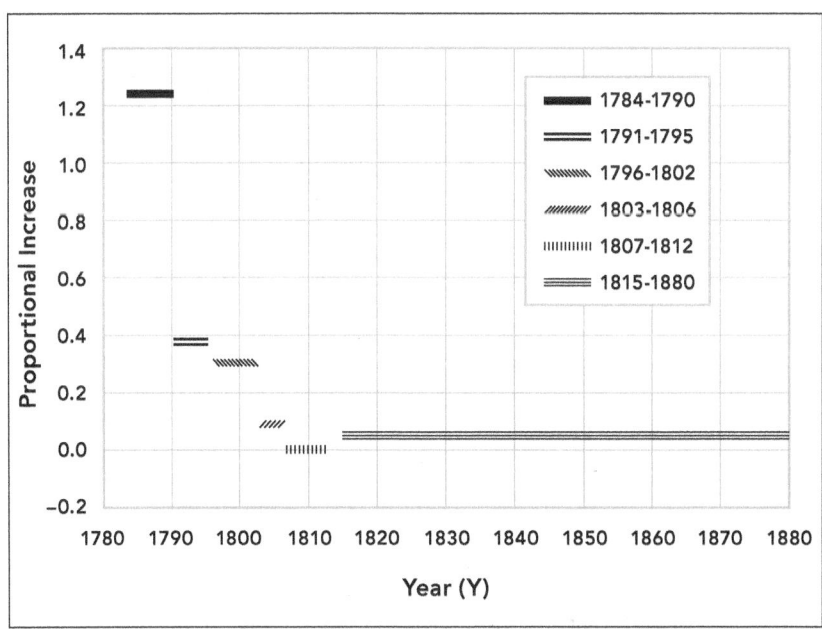

Graph 5-7. The proportional increase of growth (P) actually decreased while Whitney and Miller forced the cotton industry to pay royalties for the manufacture or use of saw-toothed gins.

that the increase in cotton production (as shown by the shipment data) was directly related to the increased number of factories established for manufacturing saw gins, on plantations and in the cities, in both the South and the North.

We found that the troubled years correspond almost exactly with the M & W Company's efforts to file patent infringement suits against cotton gin manufacturers and owners of gins. These lawsuits harassed the manufacturers of saw gins and the cotton ginners between 1794 and 1810 and lowered the yearly increase in the growth of the cotton gin manufacturing industry.

During the antebellum period, the South became the major cotton producer of the world. During the 1840s, the value of the cotton exports from the United States was nearly 70 percent of the total American cash exports. Cotton production in the United States continued to increase to the point that 85 percent of the world's exported cotton crop in 1850 came from this country.[103]

To calculate the number of early cotton gins in the South, we started with total production of cotton for the year 1796, which was 20,921 bales, approximately 10,460,000 pounds of cotton. If every ginner ginned 10,000 pounds of clean cotton lint from 30,000 pounds of green-seeded cotton picked from 100 acres (that's 100 pounds per day, the capacity of Whitney's gins), there would have been nearly 1,046 gins ginning cotton in the South in 1796. If we used the ginning data from the Kincaid gins, it would have taken 499 ginners to produce the amount of cotton ginned in 1796.

There is no way that Whitney could have manufactured 1,046 gins of 100-pounds-per-day capacity and shipped them South. However, local blacksmiths could have fabricated that quantity of gins by 1796, the year after Whitney began manufacturing gins in New Haven, Connecticut. In addition, the M & W Company became bankrupt in less than three years of manufacturing batch-type, bent-needle cotton gins.

CHAPTER 6

Hodgen Holmes' Impact on Cotton Production

The exponential increase in cotton production began in 1784 with Hodgen Holmes' invention of a practical, continuous-flow, rip-saw-toothed cotton gin. Eli Whitney, who in 1800 argued in court that the saw gin was an improvement of his bent-needle cotton gin, called these "pirated gins." These "pirated gins" fulfilled the need for expanding production of cotton. This was at least ten years before Whitney began manufacturing his bent-needle gin.

Holmes and other blacksmiths increased cotton ginning capacity each year to match the increased production of cotton. Many gin manufacturers claimed in court that they had invented new cotton gins, but Holmes' gin contained more of the features found in modern cotton gins than Whitney's inferior impractical design. Modern cotton gins follow Holmes' saw design, and the continuous-flow, rip-saw-toothed gins have cleaned most of the world's upland green-seeded cotton since the mid-1780s.

Aware of the superiority of the saw-toothed design, Phineas Miller modified Whitney's gins after they were shipped south by

Marketing cotton

A street scene of cotton producers marketing their product in Greenwood, South Carolina, in 1896. *Library of Congress*

removing the wooden cylinder bristling with bent needles and installing a cylinder equipped with a set of spaced circular saw blades. Nevertheless, these modified saw gins, because of the shape of their hoppers and the location of the breastwork, were still batch-type gins.

Whitney and Miller were opportunists. They tried to keep others from manufacturing Holmes' saw gin design while they worked the legal system to incorporate features of Holmes' design into Whitney's patent claim. Through its legal maneuvers, the M & W Company materially slowed the growth of the green-seeded cotton industry and the growth of the gin manufacturing industry. In addition, the M & W Company's monopolistic practices turned the cotton planters against the company. The company continued manufacturing its impractical design for three more years and went broke manufacturing cotton gins.

Whitney began manufacturing muskets in 1798, but he continued manufacturing and shipping bent-needle cotton gins to Miller, and Miller continued to retrofit these gins with circular saw blades.[104] In 1823, two years before his death, Whitney fabricated a saw gin for display and to promote his legacy as inventor of the cotton gin. This final saw gin was still an inferior batch-type design.

It is a tragedy for the South that Whitney never understood the advantages of Holmes' gin. He did not apply his genius for manufacturing to the manufacture of saw-toothed gins. He continued to manufacture batch-type gins, and he and Miller went bankrupt protecting their inferior design. If Whitney had fully adopted Holmes' continuous-flow, rip-saw-toothed design, the South would have been equipped with well-designed cotton gins with interchangeable parts. These machines would have had a long life, and they would have been repairable. As it was, until about 1820, the South relied on local blacksmiths to fabricate gins that, for all their utility, were often of crude construction and had a short life.

Because the M & W Company had alienated the planters with its monopolistic practices and endless court battles, the company sold few gins. Consequently, other New England companies began manufacturing cotton gins. Even so, at least 80 percent of the cotton gins used were manufactured in the South.[105] In fact, Professor Fred Amran and engineer Corle F. Pride of Minnesota believe that Catharine Greene was the true inventor of Whitney's cotton gin. It is possible that she knew about churkha gins and black-seeded cotton. Some say she eventually filed her claims concerning her invention of the cotton gin in court and received royalties from Whitney.

After the development of the saw gin, nearly every southern man who could raise the cash or credit purchased a mule, a plow, and a sack of cotton seed. With slave labor, these entrepreneurs cleared the hardwood forest of the Mississippi Delta and took up the cause of King Cotton. Cotton prices rose dramatically as England and New England used the new industrial looms to turn upland, green-seeded cotton into finished goods.[106]

Short-staple cotton sold for about half the price of long-staple Sea Island cotton, but only short-staple cotton planters had the option of expanding their acreage. As associate professor

Texas cotton gin
This sketch is titled "The First Bale of the Cotton Crop" and depicts the interior of a Texas cotton gin house in the late 1800s. With more efficient cotton gins, it was easier to get more from the cotton harvested. *Authors' collection*

of history Angela Lakwete simply stated, "Quantity won over quality."[107] Because the saw gin reduced the staple length, time-consuming roller gins continued to be used for ginning long-staple cotton. Tailored into luxurious clothing, Sea Island cotton commands premium prices, even to this day.

Holmes' saw gins had profound political and economic effects on the nation. Because slaves could clean only about one pound of cotton per day from three pounds of green-seeded raw cotton, their work did not pay for their subsistence. Hand-ginning was so uneconomical that southern planters were even considering emancipating their slaves. However, after the invention of the cotton gin, the value of healthy slaves and cotton land rapidly

"Modern spinner"

A look at a cotton-spinning machine at Mollahan Mill in Newberry, South Carolina, in 1908. As these machines became more commonplace in cotton mills, production of cotton fabrics increased.
Library of Congress

increased, and cotton gins made fortunes for the planters. Donald Holley reported that southern plantation owners of the antebellum period had one of the highest standards of living in the world.[108]

The invention of the rip-saw-toothed cotton gin also determined the future course of the nation. Increasingly dependent on a single high-value crop, the agricultural South became separate from, and yet dependent on, the industrial North. This complex relationship between North and South, which included the morality and politics of slavery, eventually led to the Civil War.

Charles Abel Bennett graciously summed up his study of the cotton gin and tried to answer the question "Who really invented the cotton gin?" Of Whitney and Holmes, he stated, "Neither received financial gain, but both should be placed in the Hall of Immortals."[109] We believe that Hodgen Holmes is the **rightful** grandfather of the modern cotton gin.

Holmes was the first to file for a caveat of invention with the War Office for a rip-saw-toothed gin in 1787. He received a caveat of invention from the War Office on March 14, 1789. However, the U.S. Patent Office was dilatory in upgrading Holmes' original caveat of invention. It did not grant a patent to Holmes in 1790, when the U.S. Patent Office was first established, or when it was revised in 1793. The office had four years to upgrade his caveat but did nothing. Holmes was finally granted a patent for a continuous-flow, rip-saw-toothed gin by the U.S. Patent Office on May 12, 1796. He was among the first to manufacture saw gins.

In addition, we think that South Carolina folklore is correct. Hodgen Holmes was the first to conceive of and fabricate a practical continuous-flow cotton gin equipped with circular rip-saw-toothed blades.

We should be grateful to Eli Whitney for his real accomplishment, which was the invention of mass production based on the production of interchangeable parts. We should also be grateful to Hodgen Holmes for the invention of the continuous-flow, rip-saw-toothed cotton gin, which is still used today, for ginning approximately 18 million bales of cotton grown in the United States—and the millions more bales grown around the world—every year.

Appendix 1: Holmes' 1796 Patent for a Rip-Saw-Toothed Cotton Gin

Title page of the patent granted to Hodgen Holmes on May 5, 1796, for a rip-saw-toothed cotton gin (on the following page):

The United States of America.

To all to whom these Letters Patent shall come:

WHEREAS *Hodgen Holmes* a citizen of the State of *Georgia*, in the United States, hath alleged that he has invented a new and useful improvement, *to wit New Machinery called the Cotton Gin*

which improvement has not been known or used before his application; has made oath, that he does verily believe that he is the true inventor or discoverer of the said improvement; has paid into the Treasury of the United States, the sum of thirty dollars, delivered a receipt for the same, and presented a petition to the Secretary of State, signifying a desire of obtaining an exclusive property in the said improvement, and praying that a patent may be granted for that purpose: THESE ARE THEREFORE to grant, according to law, to the said *Hodgen Holmes* his heirs, administrators or assigns, for the term of fourteen years, the full and exclusive right and liberty of making, constructing, using, and vending to others to be used the said improvement, a description whereof is given in the words of the said *Hodgen Holmes* himself, in the schedule hereto annexed, and is made a part of these presents.

IN TESTIMONY WHEREOF, *I have caused these Letters to be made Patent, and the Seal of the United States to be hereunto affixed.*

Given under my hand, at the City of Philadelphia this twelfth day of May, in the Year of our Lord, one thousand seven hundred and ninety six and of the Independence of the United States of America the twentieth.

G. Washington

By the President

Timothy Pickering *Secretary of State*

City of Philadelphia, TO WIT:

I DO HEREBY CERTIFY, That the foregoing Letters Patent, were delivered to me on the twelfth day of May in the year of our Lord one thousand seven hundred and ninety six to be examined; that I have examined the same, and find them conformable to law. And I do hereby return the same to the Secretary of State, within fifteen days from the date aforesaid, to wit: On this twelfth day of may in the year aforesaid. Charles Lee Attorney General

Courtesy of Robin Kaller of Kaller Historical Documents, Inc.

Transcription of the title page of Holmes' 1796 patent:

The United States of America.
To all whom these Letters Patent shall come:
Whereas, Hodgen Holmes, a citizen of the State of Georgia, in the United States, hath alleged that he has invented a new and useful improvement, to-wit, new machinery called the cotton gin, which improvement has not been known, or used before his application, has made oath that he does verily believe that he is the inventor or discoverer of the said improvement, has paid into the Treasury of the United States, the sum of thirty dollars, delivered a receipt for the same, and presented a petition to the Secretary of State, signifying a desire of obtaining an exclusive property in the said improvement, and praying that a Patent may be granted for that purpose: These are therefore to grant according to law, to the said Hodgen Holmes, his heirs, his administrators, or assigns, for the term of fourteen years, from the nineteenth day of the month of April last past, the full and exclusive rights and liberty of making, constructing, using and vending to others to be used the said improvement, a description whereof is given in the words of the said Hodgen Holmes himself the schedule hereto annexed and is made a part of these presents.

In testimony whereof, I have caused these letters to be made Patent and the seal of the United States to be hereunto affixed.

Given under my hand at the city of Philadelphia, this twelfth day of May in the year of our Lord, one thousand, seven hundred and ninety-six and of the independence of the United States of America the twentieth.

George Washington.
By the President,

Timothy Pickering,

Secretary of State

City of Philadelphia, to-wit:

I do hereby certify that the foregoing letters Patent were delivered to me, on the twelfth day of May, in the year of our Lord, one thousand, seven hundred and ninety-six to be examined, that I examined the same, and find them conformable to the law, and I do hereby return the same to the Secretary of State within fifteen days from the date aforesaid, to-wit; on this twelfth day of May, in the year aforesaid.

Charles Lee,

Attorney General

Here is a description of the saw gin in Holmes' 1796 patent:

This schedule referred to in these Letters Patent and making part of the same containing a description in the words of the said Hodgen Holmes himself of an improvement, to-wit: new machinery called the cotton gin.

Reference No. 2.

Explanation of the Whole Machinery.

This machinery for cleaning cotton from the seed can be used in the following manner, viz. The machine (standing on the floor) is made of wood, six feet, six inches wide, five feet long and five feet high, by putting this machine in motion for use of the before mentioned purpose, is to be done by the following directions:

The cylinder from fourteen inches in diameter, and six feet long with one row of teeth, to one inch, which runs on to iron gudgeons, the feeder from eight to twelve inches diameter, with two rows of wire of one inch, and six feet long and runs on two iron gudgeons, the brush from seven to twelve inches in diameter, and six feet long with two gudgeons to each cylinder, from three-quarters of an inch to one inch thick.

Hodgen Holmes

Teste, W. Urquhart, Seaborn Jones.

Department of State, to-wit:

I hereby certify that the forgoing a Letters Patent from the United States to Hodgen Holmes are a true copy of the original on record in this Department.

Given under my hand and seal of office the twenty-day of October 1797.

(Seal) Timothy Pickering

Appendix 2: Timeline of Whitney's Patent Infringement Litigation

Eli Whitney pursued multiple claims simultaneously, many of the cases lasted for several years, and the cases tended to overlay. As a result, the chronology can be difficult to follow.

April 1793: Whitney completes a full-size prototype of a bent-needle, batch-type cotton gin.

May 27, 1793: Whitney signs a partnership agreement with Phineas Miller, forming the M & W Company. Whitney will manufacture gins in New Haven, Connecticut, and Miller will lease gins and operate cotton ginneries.

May 30, 1793: Miller, in Georgia, urges Whitney, in Connecticut, to hurry the patent application process, as other parties are pursuing the manufacture of cotton gins.

June 20, 1793: Eli Whitney discusses his gin and the patent application process with Thomas Jefferson.

November 6, 1793: Whitney's patent specifications and drawings are received at the U.S. Patent Office. Whitney then begins fabricating the required patent model.

November 16, 1793: Jefferson writes Whitney that he has received the patent application. He reminds Whitney to submit a working model of the gin before a patent can be granted.

February 22, 1794: Whitney shows his model gin to his former tutor, Mr. Stiles.

March 14, 1794: Whitney presents the model gin, along with the patent documentation, to the patent committee. He is granted a patent for a bent-needle cotton gin.

Spring of 1794: William Kennedy advertises for the employment of 20 ten-year-old Negro boys for "cotton making." This could mean operating cotton gins. William Kennedy also advertises for the employment of a few Negro men to operate foot (that is, churkha) cotton gins.

March 6, 1794: Miller advertises in Georgia newspapers that plantation owners should begin planting green-seeded cotton in time for the fall harvest and that the M & W Company will gin all green-seeded cotton brought to their newly patented gins.

May 11, 1794: Whitney ships six new cotton gins to Savannah, Georgia. Two water-powered gins are installed at the Mulberry Grove plantation. The other four are installed on plantations elsewhere in Georgia and South Carolina.

June 28, 1794: Miller requests Secretary of State Edmond Randolph to delay the processing of any patent applications submitted for improvements to the cotton gin.

May 1, 1795: In newspaper advertisements, the M & W Company warns other cotton gin manufacturers of the risk of patent infringement. The M & W Company requests that all cotton gins be turned in to the M & W Company. offices, promising that those who comply will not be sued.

March 11, 1795: During the summer, fall, and winter of 1794 and 1795, Whitney builds a factory and manufactures twenty gins. The building and its contents burned on March 11, 1795. Whitney borrows money, constructs a new building, and manufactures thirty gins. He ships them south in time for the fall harvest in 1795.

September 11, 1795: A Natchez, Mississippi, newspaper reports that John Barkley's saw gin has a capacity of 750 pound of lint per day. Barkley never submits a patent application.

1795: The M & W Company files its first patent infringement suit against William Kennedy. The jury returns a verdict for the defendant on the grounds that Kennedy broke no laws. The judge advises the plaintiffs, the M & W Company, that, according to the law, the defendant had to be accused of devising, making, using, and vending a cotton gin. Kennedy could only be accused of ginning cotton; he had not made or sold a gin.

A second suit is filed against Edward Lyon for manufacturing the gin that Kennedy used. In this suit, the jury returns a non-suit verdict and requires the M & W Company to pay court costs. The

M & W Company reopens the case in 1798. In the first twenty-two lawsuits, the courts return fourteen non-suit verdicts, two defendant verdicts, four non-served, and two dismissed rulings.

November 7, 1795: Newspapers publish complaints from British spinners and weavers about damaged, knotted fibers caused by cotton gins.

1796: Southern planters seek out ginners and contract with local blacksmiths to fabricate saw gins, while increasing the acreage of green-seeded cotton that they plant each year.

May 12, 1796: Hodgen Holmes is granted a patent for a rip-saw-toothed gin.

February 15, 1797: Miller informs Whitney that a Ralph Holmes has patented a "ratchet wheel" gin. This gin contains circular saw blades whose shape is reminiscent of the ratchet wheel on the take-up roll of a handloom. *Note: The authors believe that Miller was confused about Holmes' first name.*

February 15, 1797: Miller requests that Whitney depose their friends Goodrich and Stebbins "on the subject of ratchet wheels." Unwilling to perjure themselves, Goodrich and Stebbins refuse to comply. Whitney later testifies in court that he knew about the use of circular saw blades as used in rip-saw-toothed cotton gins, but he insists that he never considered manufacturing saw gins.

1798: Whitney receives a contract for the manufacture of ten thousand stands of arms for the U.S. military. He temporarily ceases the manufacturing of cotton gins.

February 1799: Miller studies the Revised Patent Act of 1793, which specifies that anyone who discovers an improvement in a patented machine can only patent the improvement and cannot "make, use, or vent the whole machine." Holmes therefore acts legally in patenting his improved saw gin, but Miller acts illegally in retrofitting Whitney's bent-needle gins with saw-toothed cylinders.

Miller also finds a way to acquire legal ownership of Holmes' rip-saw-toothed ginning mechanism. The Revised Patent Act of 1793 states that "simply changing the form or proportion of the improvement is not inventing." Miller must convince a court that the bent-needle ginning mechanism is the same as the rip-saw-toothed ginning mechanism. If he is successful, then Holmes' 1796 patent will be nullified.

1799: Whitney prepares for future trials by fabricating a model cotton gin bristling with twelve rows of bent-needle teeth. On the same rotating shaft, he assembles six circular saw blades spaced to operate in the slotted tangential breastwork. Numerous lawsuits are pursued in 1798 to 1801, and this misleading model is used in courtroom demonstrations.

June 20, 1800: Holmes is named as a party in a lawsuit. Judge Joseph Clay issues a Rule Absolute requiring Holmes to show the court why his patent should not be nullified. Holmes is unsuccessful.

April 1801: Whitney persuades the U.S. Congress to change the Patent Law. Defendants can be accused of patent infringement if they "make, devise, use, or vent" a cotton gin.

September 6, 1802: Judge William Stephens rules that the two ginning mechanisms, bent-needle and rip-saw-toothed, are essentially the same. The judge nullifies Holmes' 1796 patent.

1801 to 1803: Whitney sells State Manufacturing Licenses to South Carolina, Georgia, and Tennessee for a total of $90,000. These blacksmiths and manufacturers in these states can now manufacture cotton gins royalty free. Whitney's proceeds go toward paying his legal bills.

1803: Holmes purchases a license from the M & W Company to continue manufacturing saw gins.

1803 to 1807: Miller and Whitney file five more patent infringement suits. In preparation for the trials, Whitney hires a draftsman to duplicate the drawings from his 1793 patent and from Holmes' 1796 patent. He asks Secretary of State James Madison to compare the originals with the duplicates and certify the accuracy of the copies. The certified copies are used in trials in 1804, 1806, 1807, and 1810.

1803: Whitney sues Arthur Fort and John Powell for using saw gins, the patent for which he owns. The defendants deny the charge, claiming that since Whitney did not invent the saw gin, they could not have infringed. Whitney insists that his patent applies to both bent-needle gins and rip-saw-toothed gins. He falsely testifies that Goodrich and Stebbins will corroborate that he discussed saw blades with them.

December 19, 1806: The court renders verdicts for the plaintiff. In addition, Whitney requests a perpetual injunction against Holmes' estate for the manufacture of saw gins. The judge complies, issuing the following statement:

> *"A Mr. Holmes has cut teeth in plates of iron, and passed them over the cylinder. This is certainly a meritorious improvement in the mechanical process of constructing this machine. But at last, what does it amount to, except a more convenient mode of making the same thing? Every characteristic of Mr. Whitney's machine is preserved. Mr. Whitney may not be at liberty to use Mr. Holmes' iron plates, but certainly Mr. Holmes' improvement does not destroy Mr. Whitney's patent rights. Let the decree for a perpetual injunction be entered."*

1807: Judge William Johnson upholds William Stephens' 1802 decision of nullifying Holmes' patent, calling Whitney's patent "the basic cotton gin patent." He declares Holmes' rip-saw-toothed gin merely an adaptation of the bent-needle gin that is covered by Whitney's patent.

In Whitney vs. Carter, defense witnesses testify that they had seen circular saw blades, protected by ribs, used for removing coconut husks in the Caribbean and for deburring wool in Ireland. They state that Holmes may have reinvented the circular saw blade. Nevertheless, the U.S. District Court finds that Whitney's patent rights for the bent-needle gin are senior to Holmes' saw gin.

1807 and 1811: Southern senators defeat Whitney's request to the U.S. Congress to extend his patent.

May 10, 1808: Plaintiff verdicts are issued against Isaiah Carter for $2,000 and against Charles Gachet for $1,500, the equivalent of triple royalty fines.

1809: Whitney fulfills a contract for the manufacture of ten thousand stands of arms, ten years later than promised.

1823: Whitney fabricates a batch-type model gin that incorporates circular saw blades and a tangential breastwork. This model proves that Whitney never understood the operating principles of the continuous flow, rip-saw-toothed cotton gin.

1963: Bennett analyzes the Holmes and Whitney patents. Because Whitney's patent was granted before Holmes' patent, Bennett agrees that Whitney was the senior inventor. However, Bennett also states that Holmes' gin was the superior design.

Appendix 3: A Historical Landmark of Agricultural Engineering

In 1986, the Historical Landmark Committee of the American Society of Agricultural Engineers (ASAE) erected a bronze plaque in Savannah, Georgia, to commemorate the invention of the cotton gin. Committee chairman Arnold Skromme and committee member William D. Mayfield evaluated the historical information found by the committee and approved the following text, which names Eli Whitney as the inventor of a spike-toothed cotton gin and Hodgen Holmes as the inventor of an improved saw-toothed cotton gin. This plaque may be viewed today in Savannah.

The concrete foundation of Whitney's cotton gins, installed on the Mulberry Grove plantation in 1794, can still be seen just off Georgia Highway 17, about 12 miles north of Savannah. We have used the word *needle* because currently, in modern times, a spike is the name for a very large nail.

The Invention of the Cotton Gin
A Historical Landmark
of Agricultural Engineering

This creative development, which was responsible for the survival of the cotton industry in the United States, occurred on General Nathanael Greene's plantation near Savannah, 10 miles northeast of this marker. Separation by hand labor of the lint from the seed of the desired upland variety of cotton produced only one pound per day per person.

Eli Whitney, a native of Massachusetts and Yale Law Graduate, came to Georgia to teach school in late 1792, at age 27. Mrs. Catharine Greene, widow of General Greene, invited Whitney to her plantation and urged him to design a cotton gin. He secluded himself for 10 days in the spring of 1793, with a basket of cotton bolls. He discovered that a hooked wire could pull the lint through a slot in the basket, leaving the seeds inside. In his patent application, Whitney described the process as: consisting of spikes driven into a wooden cylinder and having a slotted bar through which these spikes passed and having a brush to clean the spikes. The result was a hand-operated cotton gin that produced over 50 pounds per person per day. It was patented March 14, 1794.

Henry Ogden Holmes of Georgia, a resourceful, practical mechanic on the Kincaid plantation of Fairfield County, South Carolina, invented an improved gin and was granted a patent on May 12, 1796. His continuous-flow gin used teeth on a circular steel blade that passed through spaces between ribs. The circular rip-saw-tooth gin with improvements, capable of ginning 1,000s of pounds per day, was still in use in 1985.

Officials of the Cotton Exchange Commission Building, which faces this marker, shipped from the Port of Savannah thousands of bales to a new worldwide industry and brought prosperity to the South.

Dedicated by the American Society of Agricultural Engineers

July 1986

Appendix 4: The Data Analysis Referenced in Chapter 5

In Chapter 5 we discussed the growth of the early southern cotton industry. In many ways, it is analogous to the growth rate of cells in a petri dish. Cellular growth has a continuous proportional increase of growth (called P) over the previous growth period until food scarcity or some external force interferes with growth. The new value of P is then made up of the two years added together.

The data shown on pages 96-97 in Chapter 5 are shown in Graph A4-1 on the following page.

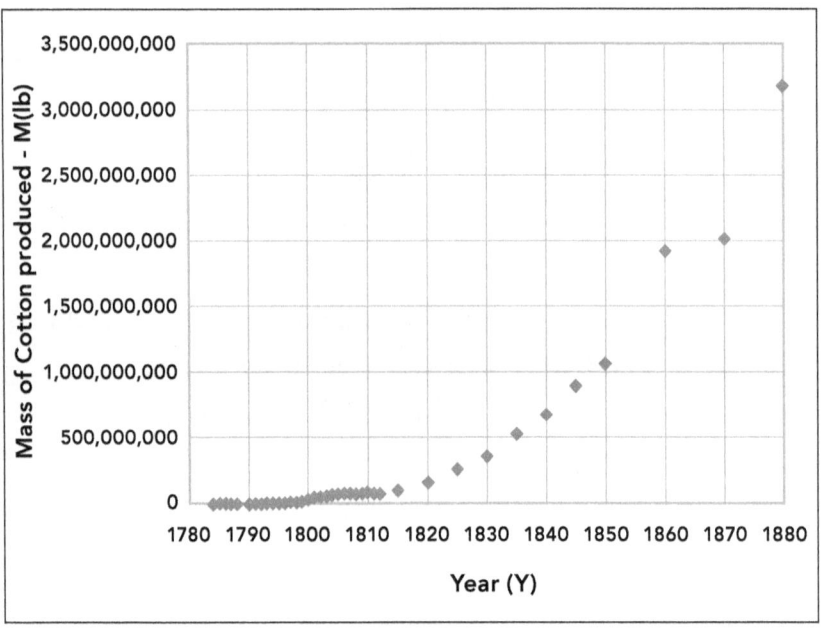

Graph A4-1. Annual shipment of cotton lint from the southern United States, 1784 to 1880. Both the vertical and horizontal axes have linear (Cartesian) scale.

We further examined these data by plotting the mass of cotton produced on a vertical semi-logarithmic scale, as shown in the graph below.

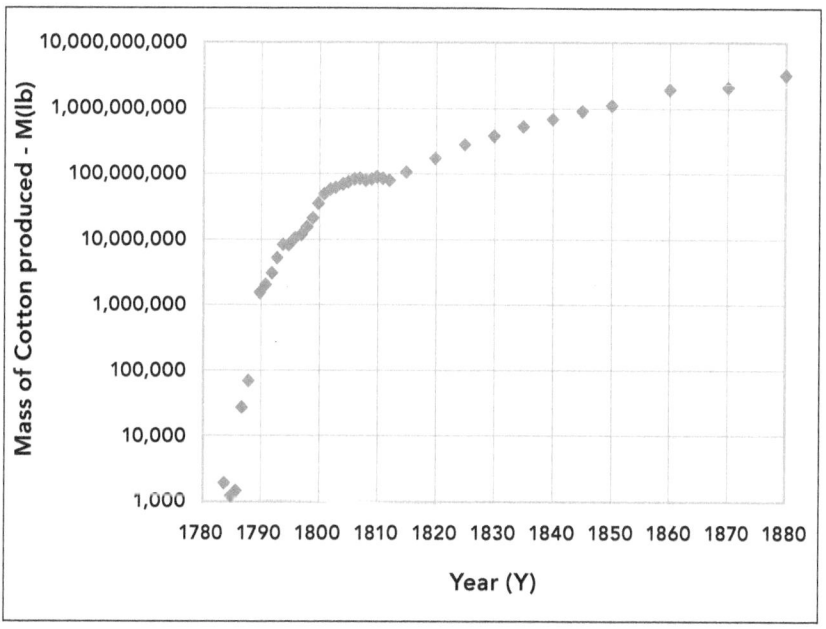

Graph A4-2. Annual shipment of cotton lint from the southern United States, 1784 to 1880. Mass of cotton plotted on the vertical axis with a logarithmic scale.

Considering the data on pages 96-97 in Chapter 5 as well as the presentation of these data in Graph A4-2, we concluded that the cotton production timeline could be divided into six segments (major periods) so we could study what happened during each of the periods. A typical segment from 1784 to 1790, the first time segment plot, is shown in Graph A4-3.

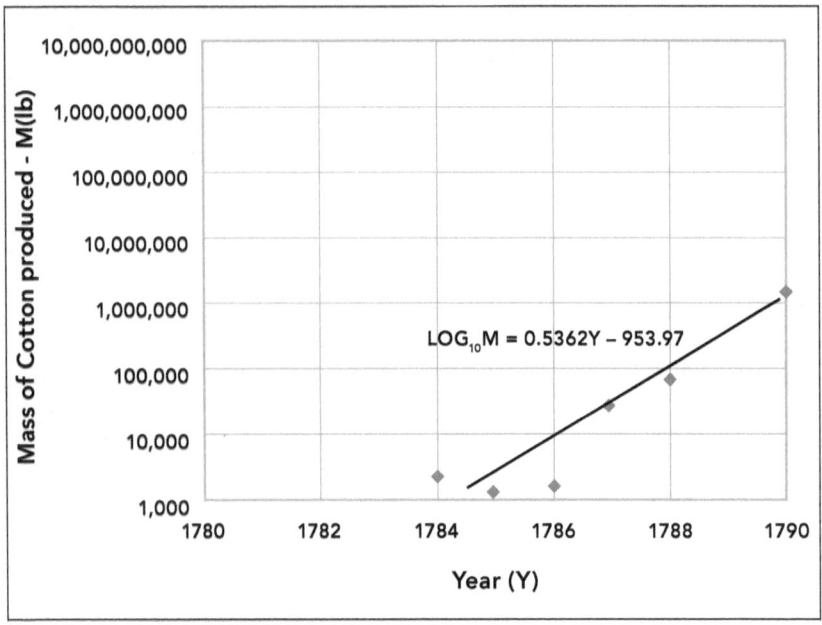

Graph A4-3. Annual shipment of cotton lint from the southern United States, 1784 to 1790 (plotted on semi-log scale).

We fit the following equation to each of the six major time segments:

Equation A4-1.

$$\log_{10} M = a \times Y + b$$

Equation A4-1a.

$$\therefore M = 10^{aY + b}$$

where:

M = mass of cotton produced (lb)

Y = year

a and b are both constant within each of the six time segments.

The rate of change of the mass of cotton shipped (in pounds per year) is:

Equation A4-2.

$$\frac{dM}{dY} = a \times \ln(10) \times 10^{aY + b}$$

The rate of change of the mass of cotton shipped divided by the mass of cotton produced gives the proportional increase (P) in cotton produced relative to the cotton produced during the prior time segment ([pounds/year] per pound):

Equation A4-3.

$$P = \frac{\left(\frac{dM}{dY}\right)}{M} = \frac{a \times \ln(10) \times 10^{aY + b}}{10^{aY + b}} = a \times \ln(10)$$

A proportional increase in clean cotton production (P) would depend on the rate of increase in the capacity and the number of saw-toothed cotton gins. Therefore, the decreasing slope of each line segment over time indicates that there were problems with the production of seed cotton or with the manufacture of cotton saw gins. We postulated that the proportional increase in cotton production (as shown by the shipment data) was directly related to the increased number of factories established for manufacturing saw-toothed gins, on plantations and in the cities, in both the South and the North. We found that the troubled years correspond almost exactly with the M & W Company's efforts to file patent infringement suits against cotton gin manufacturers and owners of gins. These lawsuits harassed the manufacturers of saw-toothed gins and the cotton ginners between 1794 and 1810 and lowered the proportional increase in the growth (P) of the cotton gin manufacturing industry.

Endnotes

Chapter 1

1   C. A. Bennett, *Saw and Toothed Cotton Ginning Developments* (Washington, D.C.: USDA, 1963).

2   A. Lakwete, *Inventing the Cotton Gin* (Baltimore, MD: Johns Hopkins University Press, 2003).

3   Bennett, *Saw and Toothed Cotton Ginning Developments.*

4   G. Healy-Johnson, *Cotton* (Mankato, MN: Creative Education, 2000).

5   H. Thompson, *The Age of Invention: A Chronicle of Mechanical Conquest* (Fairbanks, AK: Project Gutenberg Literary Foundation, originally published in 1921, available at www.gutenberg.org/ebooks/2900).

6   C. M. Green, *Pioneer Inventor* (New York: Fawcett Publications, 1965).

7   Lakwete, *Inventing the Cotton Gin.*

8   Ibid.

9   Ibid.

10 Ibid.

11 M. C. McMillan, *The Manufacture of Cotton Gins: A Southern Industry, 1793–1860* (Dallas: Cotton Gin and Oil Press, 1993).

12 W. Mayfield, R. Baker, S. E. Hughes, and A. C. Griffin, *Ginning Cotton to Preserve Quality* (Washington, D.C.: USDA, 1983).

13 J. W. Smith, *No Holier Spot of Ground* (Huntsville: Sam Houston State University, Texas Review Press, 2004).

14 D. Holley, *The Second Great Emancipation: The Mechanical Cotton Picker, Black Migration, and How They Shaped the Modern South* (Fayetteville: University of Arkansas Press, 2000).

15 M. B. Hammond, "Correspondence of Eli Whitney Relative to the Invention of the Cotton Gin," *American Historical Review* 3 (October 1897): 90–127.

16 R. D. Porcher and S. Fick, *The Story of Sea Island Cotton* (Layton, UT: Wyrick and Co., 2005).

17 G. R. Merrill, A. R. Macormac, and H. R. Maurersberer, *American Cotton Handbook: A Practical Reference Book for the Entire Cotton Industry* (New York: American Cotton Handbook Co., 1949).

18 H. Evans, G. Buckland, and D. Lefer, *They Made America* (New York: Little, Brown, and Co., 2004).

19 R. L. Haney, *Milestones: Marking Ten Decades of Research* (College Station: Texas Agricultural Experiment Station, 1989).

20 K. G. Britton, *Bale o' Cotton: The Mechanical Art of Cotton Ginning* (College Station: Texas A&M University Press, 1992).

21 F. L. Lewton, *Historical Notes on the Cotton Gin* (Washington, D.C.: Smithsonian Institution, 1938).

22 P. A. Jones and E. N. Simons, *The Story of the Saw* (London: Spear and Jackson, 1961).

23   Thompson, *The Age of Invention: A Chronicle of Mechanical Conquest.*

24   Lakwete, *Inventing the Cotton Gin.*

25   E. H. Cameron, *Samuel Slater* (Freeport, ME: Bond Wheelwright Co, 1960).

26   J. W. West, *King Cotton: Fiber of Slavery* (Slavery in America, 2006, available at: www.slaveryinamerica.org/history/hs_esw_cotton.htm).

27   Cameron, *Samuel Slater.*

28   W. Kaempffert, ed., *A Popular History of American Inventions*, Vol. II (New York: Charles Scribner, 1924).

29   Lakwete, *Inventing the Cotton Gin.*

30   J. A. B. Scherer, *Cotton as a World Power: A Study in the Economic Interpretation of History* (New York: Frederick A. Stokes Co, 1916).

31   J. Mirsky and A. Nevins, *The World of Eli Whitney* (New York: Macmillan, 1952).

32   Ibid.

33   Ibid.

34   S. Bruchey, *Cotton and the Growth of the American Economy: 1790–1860* (New York: Harcourt Brace and World, 1967).

35   Mirsky and Nevins, *The World of Eli Whitney.*

36   Evans, Buckland, and Lefer, *They Made America.*

37   Lakwete, *Inventing the Cotton Gin.*

38   Ibid.

Chapter 2

39 D. Olmsted, *Memoir of Eli Whitney, Esq.* (New Haven, CT: Durrie & Peck, 1832).

40 Jones and Simons, *The Story of the Saw.*

41 F. Johnson, *Fairfield Family Histories (1700-1982)* (Fairfield, SC: Fairfield Publishers, 1984).

42 A. B. Walker, *History of the Kincaid Family and the First Cotton Gin.* (Unpublished, 1964.) Copies are located in the Clemson University Library (Clemson, SC) and the Winnsboro City Library (Winnsboro, SC).

43 Lakwete, *Inventing the Cotton Gin.*

44 Walker, *History of the Kincaid Family and the First Cotton Gin.*

45 D. A. Tompkins, *Cotton and Cotton Oil*, 2 Volumes (Dallas: Texas Cotton Ginners' Association, 1901).

46 Lakwete, *Inventing the Cotton Gin.*

47 Ibid.

48 Tompkins, *Cotton and Cotton Oil.*

49 Ibid.

50 McMillan, *The Manufacture of Cotton Gins: A Southern Industry, 1793–1860.*

51 K. Hutchinson, Letter to Miss Janie Hutchinson, dated June 12, 1937.

52 Lakwete, *Inventing the Cotton Gin.*

Chapter 3

53 Hammond, "Correspondence of Eli Whitney Relative to the Invention of the Cotton Gin."

54   Ibid.

55   Britton, *Bale o' Cotton.*

56   Evans, Buckland, and Lefer, *They Made America.*

57   T. Sutton and D. K. Utley, *From Can See to Can't: Texas Cotton Farmers on the Southern Prairies* (Austin: University of Texas Press, 1997).

58   B. Norman, *Inventing America* (New York: Taplinger Publishing, 1976).

59   K. Chowder, "Eureka," *Smithsonian Magazine* (September 2003).

60   E. Whitney, Copy of original letters patent, (Savannah, GA: Federal Circuit Courthouse, 1793).

61   Mirsky and Nevins, *The World of Eli Whitney.*

62   Porcher and Fick, *The Story of Sea Island Cotton.* (Charleston, S.C.: Wyrick and Co., 2005).

63   Merrill, Macormac, and Maurersberer, *American Cotton Handbook* (New York: American Cotton Handbook Co., 1949).

64   Kaempffert, *A Popular History of American Inventions*, Vol. II (New York: Charles Scribner, 1924).

65   Bennett, *Saw and Toothed Cotton Ginning Developments.*

66   Lewton, *Historical Notes on the Cotton Gin* (Washington, D.C.: Smithsonian Institution, 1937).

67   Ibid.

68   Bennett, *Saw and Toothed Cotton Ginning Developments*; G. Bathe and D. Bathe, *Oliver Evans: A Chronicle of Early American Engineering* (Philadelphia: Historical Society of Pennsylvania, 1935).

69   Evans, Buckland, and Lefer, *They Made America.*

70   Tompkins, *Cotton and Cotton Oil.*

71   Scherer, *Cotton as a World Power.*

Chapter 4

72   Lakwete, *Inventing the Cotton Gin.*

73   D. H. Bacot, "Rise of the Cotton Industry," Charleston, South Carolina's *The News and Courier*, 1939.

74   Lakwete, *Inventing the Cotton Gin.*

75   Ibid.

76   J. S. Bolick, *A Fairfield Sketchbook* (Clinton, SC: Jacobs Press, 2000).

77   Scherer, *Cotton as a World Power.*

78   C. Roberts, *Founding Mothers: The Women Who Raised Our Nation* (New York: Harper Collins, 2004).

79   Bennett, *Saw and Toothed Cotton Ginning Developments.*

80   Lakwete, *Inventing the Cotton Gin.*

81   Ibid.

82   Lewton, *Historical Notes on the Cotton Gin*; H. Rabb, *Biographical Sketches of the Kincaid, McMorries, Watt, Glazer, and Rabb Families*, 1936.

83   Tompkins, *Cotton and Cotton Oil.*

84   T. G. Fessenden, *An Essay on the Law of New Patents for New Inventions* (Boston: D. Mallory and Co., 1810).

85   O. Evans, *The Young Millwright and Miller's Guide* (Philadelphia: Carey, Lea and Blanchard, 1834); Bennett, *Saw and Toothed Cotton Ginning Developments.*

86   Bennett, *Saw and Toothed Cotton Ginning Developments.*

87   McMillan, *The Manufacture of Cotton Gins.*

88   Bennett, *Saw and Toothed Cotton Ginning Developments;* H. P. Smith, *Farm Machinery and Equipment* (New York: McGraw-Hill, 1948).

89   Bennett, *Saw and Toothed Cotton Ginning Developments.*

90   "Historical Society Finds Original Eli Whitney Cotton Gin," *New Haven Colony Historical Society Journal* 8(5), 1959.

91   Ibid.

92   Bennett, *Saw and Toothed Cotton Ginning Developments.*

93   Olmsted, *Memoir of Eli Whitney, Esq.*

94   Evans, Buckland, and Lefer, *They Made America.*

Chapter 5
95   Olmsted, *Memoir of Eli Whitney, Esq.*

96   Ibid.

97   Ibid.

98   Bruchey, *Cotton and the Growth of the American Economy: 1790–1860.*

99   Lakwete, *Inventing the Cotton Gin.*

100   Ibid.

101   Scherer, *Cotton as a World Power.*

102   Ibid.

103   Ibid.

Chapter 6
104   Lakwete, *Inventing the Cotton Gin.*

105   McMillan, *The Manufacture of Cotton Gins.*

106   Holley, *The Second Great Emancipation.*

107   Lakwete, *Inventing the Cotton Gin.*

108   Holley, *The Second Great Emancipation: The Mechanical Cotton Picker, Black Migration, and How They Shaped the Modern South.*

109   Bennett, *Saw and Toothed Cotton Ginning Developments.*

References

All documents cited in the all chapters are listed here. Documents with asterisks (*) are not cited in the text:

*American Society of Mechanical Engineers. *Burton Farmers Gin*. New York: American Society of Mechanical Engineers, 1994.

*Aran, Mark, and Rick Wartzman. *The King of California: J. G. Boswell and the Making of a Secret American Empire*. Cambridge, MA: Perseus Book Group, 2004.

Bacot, D. H. "Rise of the Cotton Industry." Charleston, South Carolina's *The News and Courier*, 1939.

*Barlow, Ronald S. *Farm Implements and Machinery: 1630–1930*. Iola, WI: Krause Publications, 2003.

*Bates, Edward Craig. "The Story of the Cotton Gin." *New England Magazine*, reprinted by the Westborough Historical Society, 1899.

Bathe, Greyville, and Dorothy Bathe. *Oliver Evans: A Chronical of Early American Engineering*. Philadelphia: Historical Society of Pennsylvania, 1935.

*Bellis, Mary. "Inventor's Guide."
Available at: www.about.com. Accessed March 25, 2004.

*Bennett, Charles Abel. *Cotton and Toothed Cotton Ginning Developments*. Beltsville, MD: USDA Agricultural Research Service, Agricultural Engineering Research Division, 1933.

————. *Saw and Toothed Cotton Ginning Developments*. Washington, D.C.: USDA, 1963.

Bolick, Julian Stevenson. *A Fairfield Sketchbook*. Clinton, SC: Jacobs Press, 2000.

Britton, Karen G. *Bale o' Cotton: The Mechanical Art of Cotton Ginning*. College Station: Texas A&M University Press, 1992.

Bruchey, Stuart. *Cotton and the Growth of the American Economy: 1790–1860*. New York: Harcourt Brace and World, 1967.

*Butterworth, Benjamin. *The Growth of Industrial Arts*. Washington, D.C.: U.S. Government Printing Office, 1886.

Cameron, Edward H. *Samuel Slater*. Freeport, ME: Bond Wheelwright Co, 1960.

*de Camp, L. Sprague. *The Ancient Engineers*. New York: Ballantine Books, 1960.

*Childs, Arney R. *Calendar Kincaid-Anderson Papers, 1767–1900*. Charleston, SC: South Carolina Historical Society, 1958.

Chowder, Ken. "Eureka." *Smithsonian Magazine*, September 2003.

Evans, Harold, Gail Buckland, and David Lefer. *They Made America*. New York: Little, Brown, and Co., 2004.

Evans, Oliver. *The Young Millwright and Miller's Guide*. Philadelphia: Carey, Lea and Blanchard, 1834.

Fessenden, Thomas Green. *An Essay on the Law of New Patents for New Inventions*. Boston: D. Mallory and Co., 1810.

*Giles, Joseph, and Frances Giles. *The Ingenious Yankees*. New York: Thomas Y. Crowell, 1976.

Green, C. M. *Pioneer Inventor.* New York: Fawcett Publications, 1965.

*Grun, Bernard. *Timetable of History.* New York: Simon and Schuster, 1946.

Hammond, M. B. "Correspondence of Eli Whitney Relative to the Invention of the Cotton Gin." *American Historical Review* 3 (October 1891): 90–127.

Haney, Robert L. *Milestones: Marking Ten Decades of Research.* College Station: Texas Agricultural Experiment Station, 1989.

Healy-Johnson, Guinevere. *Cotton.* Mankato, MN: Creative Education, 2000.

"Historical Society Finds Original Eli Whitney Cotton Gin." *New Haven Colony Historical Society Journal* 8(5), 1959.

*Holcomb, Brent H. *History of Fairfield County, South Carolina.* Columbia, SC: SCMAR, 2003.

Holley, Donald. *The Second Great Emancipation: The Mechanical Cotton Picker, Black Migration, and How They Shaped the Modern South.* Fayetteville, AR: University of Arkansas Press, 2000.

*Holmes, Hodgen. Copy of original letter patent granted on May 12, 1796. Savannah, GA: Federal Circuit Courthouse, 1796.

*Hutchinson, W. H. Letter to Miss Janie Hutchinson, dated June 12, 1937.

*Jensen, James. *John Deere Cotton Harvesters.* Ankeny, IA: James. K. Jensen, 2001.

*Johnson, C. "Alien Invaders: Sad Stories." *The Furrow,* November 2003.

*———. "Legendary Lands: Hunting Cotton History." *The Furrow,* Spring 2004.

Johnson, Faye. *Fairfield Family Histories (1700–1982).* Fairfield, SC: Fairfield Publishers, 1984.

Jones, P. A., and E. N. Simons. *The Story of the Saw.* London: Spear and Jackson, 1961.

Kaempffert, Waldemar, ed. *A Popular History of American Inventions*, Vol. II. New York: Charles Scribner, 1924.

*Kendrick-Peabody, Erin. "New Approved Fungus May Help 'Clean Up' Cotton." Washington, D.C.: USDA Agricultural Research Service, 2004.

*Kincaid, James. Will dated April 12, 1802, recorded in Book 4, pages 39-40, APT 20, file 270, University of South Carolina Library, Columbia, SC.

*Kutzbach, H. D., and R. G. Quick. 1999. Part 1.6.1–1.6.7 *Harvesters and Threshers: Grain*. In *CIGR Handbook of Agricultural Engineering*, Vol. III, edited by B. A. Stout and B. Cheze. Plant Production Engineering (311–347). St. Joseph, MI: ASAE.

Lakwete, Angela. *Inventing the Cotton Gin*. Baltimore, MD: Johns Hopkins University Press, 2003.

*Leitner, Claude C. "Unknown Maker of America: Henry Ogden Holmes, the Real Inventor of the Cotton Gin." *Cotton Gin and Cotton Oil News*, 1933.

Lewton, Frederick L. *Historical Notes on the Cotton Gin*. Washington, D.C.: Smithsonian Institution, 1937.

Mayfield, William, Roy Baker, S. E. Hughes, and A. C. Griffin. *Ginning Cotton to Preserve Quality*. Washington, D.C.: USDA, 1983.

*Mayfield, William D., and W. S. Anthony. "The Development of the Cotton Gin." Accessed May 14, 2006.

McMillan, M. C. *The Manufacture of Cotton Gins: A Southern Industry, 1793–1860*. Dallas: Cotton Gin and Oil Press, 1993.

Merrill, G. R, A. R. Macormac, and H. R. Maurersberer. *American Cotton Handbook: A Practical Reference Book for the Entire Cotton Industry*. New York: American Cotton Handbook Co., 1949.

Mirsky, Jeannette, and Allan Nevins. *The World of Eli Whitney*. New York: Macmillan, 1952.

Norman, Bruce. *Inventing America*. New York: Taplinger Publishing, 1976.

Olmsted, Denison. *Memoir of Eli Whitney, Esq.* New Haven, CT: Durrie & Peck, 1832.

Porcher, Richard Dwight, and Sarah Fick. *The Story of Sea Island Cotton.* Layton, UT: Wyrick and Co., 2005.

Rabb, Horace. *Biographical Sketches of the Kincaid, McMorries, Watt, Glazer, and Rabb Families*, 1936.

*Richey, Clarence Bentley. *Fifty Years of Engineering Farm Machinery.* West Layafette, IN: Clarence B. Richey, 1989.

Roberts, Cokie. *Founding Mothers: The Women Who Raised Our Nation.* New York: Harper Collins, 2004.

*Rust, John. "The Origin and Development of the Cotton Picker." West Tennessee Historical Society Papers 7: 45–48, 1953.

Scherer, James Augustin Brown. *Cotton as a World Power: A Study in the Economic Interpretation of History.* New York: Frederick A. Stokes Co., 1916.

*Seabrook, W. B. *A Memoir on the Origin, Cultivation, and Uses of Cotton.* Charleston, SC: Agricultural Society of South Carolina, 1844.

Smith, Harris Pearson. *Farm Machinery and Equipment.* New York: McGraw-Hill, 1948.

Smith, John Warren. *No Holier Spot of Ground.* Huntsville: Sam Houston State University, Texas Review Press, 2004.

*Stefferud, Alfred, ed. *The Yearbook of Agriculture 1960: Power to Produce.* Washington, D.C.: USDA, 1960.

Sutton, Thad, and Dan K. Utley. *From Can See to Can't: Texas Cotton Farmers on the Southern Prairies.* Austin: University of Texas Press, 1997.

*Texas Agricultural Experiment Station. "Mechanical Harvesting of Cotton in Northwest Texas." TAES Circular No. 52, 1928.

Thompson, Holland. *The Age of Invention: A Chronical of Mechanical Conquest.* Fairbanks, AK: Project Gutenberg Literary Foundation, originally published in 1921.

Tompkins, Daniel Augustus. *Cotton and Cotton Oil.* 2 Vols. Dallas: Texas Cotton Ginners' Association, 1901.

Walker, Alice A. B. *History of the Kincaid Family and the First Cotton Gin.* Unpublished, 1964. Copies are located in the Clemson University Library (Clemson, SC) and the Winnsboro City Library (Winnsboro, SC).

Welch, Catherine A. *Eli Whitney.* Minneapolis: Lerner Publishing, 2007.

West, J. W. *King Cotton: Fiber of Slavery.* Slavery in America, 2006.

Whitney, Eli. Copy of original letters patent. Savannah, GA: Federal Circuit Courthouse, 1793.

*Will, Oscar III. *Wheel Loader Legacy.* Topeka, KS: Farm Collector, 2006.

*Yaha, Stephen. *Big Cotton.* New York: Viking, 2005.

INDEX

www.ingramcontent.com/pod-product-compliance
Lightning Source LLC
Chambersburg PA
CBHW020914180526
45163CB00007B/2723